HONEY BEES

Brother Adam inspecting hives at Buckfast Abbey. His apiary is arranged in groups of four hives, which face north, south, east and west.

HONEY BEES

A GUIDE TO MANAGEMENT

RON BROWN

The Crowood Press

First published in 1988 by
The Crowood Press Ltd
Ramsbury, Marlborough
Wiltshire SN8 2HR

Paperback edition 1998

British Library Cataloguing in Publication Data

Brown, Ron, *1934–*
   Honey bees: *a guide to management.*
   1. Great Britain. Livestock: Honey-bees.
   Breeding
   I. Title
   638′.1

   ISBN 1 86126 174 8

Acknowledgements
The illustration on page 7 is reproduced by permission of the British
Museum (Natural History).

Photographs by P.P. Rosenfelt (Frontispiece and page 9), and Richard Brown
(page 52). All others are by the author.

Line illustrations by Janet Sparrow.

Typeset by Alacrity Phototypesetters, Banwell Close, Weston-super-Mare
Printed in Great Britain by Redwood Books, Trowbridge, Wiltshire

Contents

Introduction

It is probably true to say that honey bees have always fascinated people in every age. Primitive man hunted them, taking honey and wax from their wild nests in caves and trees. We have the evidence in cave drawings dating back 10,000 years or more, of human figures taking combs from wild bee colonies using smoke, just as we have pictures of wild cattle being hunted at the same period. Keeping bees, as opposed to hunting them, only dates back about 4,000 years but was certainly established in Britain 1,000 years ago, and is well documented in William the Conqueror's 'Domesday Book'. Modern beekeeping with bees on moveable frames in wooden boxes came into being only about 150 years ago.

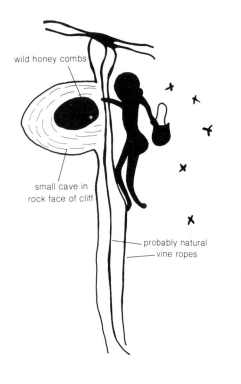

wild honey combs

small cave in
rock face of cliff

probably natural
vine ropes

Collecting wild honey, c.7000BC.
After rock painting found
in the south of Spain.

An old English skep made of willow bound with straw,
c.1550 AD, adapted from a sixteenth century woodcut.

Progress in managing bees has equalled that in managing cattle, sheep, goats and geese, and the purpose of this book is to encourage farmers and smallholders to diversify and add another dimension to their present enterprise. Bees have one very great advantage over all other stock, they forage on other people's property over a radius of three miles from their hives, and in return ask only for a few hours of our time and a few square yards of otherwise unproductive land, the edge of a copse or the corner of a field. The current movement towards conservation and the creation of small nature reserves is gathering momentum, and bees should be considered as a part of this.

In connection with bees we usually think first of honey, beeswax and other hive products, but in fact their value as pollinators of food crops is reckoned to be at least twice that of all the honey and wax produced. Fruit orchards, fields of beans and other crops, vegetable gardens and allotments, all depend on insect pollination, and by far the most important agent is the bee, the best friend of farmers and gardeners everywhere.

1 First Steps for Beginners

When the urge to keep bees first hits you, the impulse is to go out and buy a hive straight away, and learn by doing, in isolation. Or it may happen that you hear of some bees for sale and suddenly realise that you have a wonderful opportunity to do something new and quite different. This is not the best way to start; don't worry about one missed opportunity – as soon as you link up with local beekeepers you will hear of many more. The best first step is to join your local branch of the Beekeepers' Association, preferably in the autumn, and attend their programme of winter lectures, which are held frequently and cover all aspects of beekeeping. Every county in Britain has such an association, usually with several local branches, and once you have joined you will be made welcome, not only at your own branch but also at other branches. In addition to lectures, regular courses of instruction are held in different centres over the winter, and you will also meet dozens of other beekeepers, some experienced, but some who started only the previous year. They will all be happy to tell you about the problems they had and how they dealt with them.

The second step is to do some reading. Your nearest public library will help you, but even better will be the library of your local association. In Devon, for example, there are thirteen local branches and each one has its library of bee books you can borrow. Read one or two beginners' books, then buy a book and keep it handy for reference.

The third step is to take a monthly beekeeping journal. There are several in the Useful Addresses list, and they all carry topical articles, forthcoming events, advertisements from firms selling hives, veils, wax foundation and so on, as well as a column of secondhand hives and equipment for sale.

By the spring you will have a background of knowledge to help you make a much better decision on what to buy and how to start up. It may be tempting to buy a strong colony in the hope of a quick return of honey this summer, but sometimes the difficulty of handling a large stock as a beginner can put a newcomer off for life. It is far better to arrange with a local beekeeper to buy a four-frame nucleus with a young queen, taking delivery at the end of May or early in June. Regular feeding with sugar syrup will enable the nucleus to expand into a full colony on ten or eleven frames by July. They will be very quiet and easy to handle and as they grow in strength you will gain in experience, confidence and the necessary manual skills of beekeeping. There will be no swarming problems to cope with in your first year and with autumn feeding the stock will cope with the coldest winter.

By the following spring you will be well placed to increase to two stocks by making an artificial swarm from your own hive in May (*see* page 69). If you have built or acquired a couple of empty hives over the winter, you will also be ready to take or buy in swarms some time in May or June, and by July you will have at least three good stocks and the experience necessary to manage them. You should also have a crop of honey and the opportunity to learn the techniques involved in taking it off, extracting, bottling and preparing it for sale. It will still only be fifteen months since you first owned bees but you will have come a long way, and it might be wise to stay with just two or three hives for another

Ron Brown inspects a frame of brood in April (from an over-wintered nucleus).

First Steps for Beginners

year. It is much better to make your mistakes on two or three hives than on twenty. Possibly three hives are all you intend to have anyway. In either case, make a four-frame nucleus from your strongest stock some time in the summer (see page 60) and see it through the winter to sell to another beginner, to increase your own stocks or as a reserve in case anything goes wrong.

Before you get your first bees, it is necessary to choose a site and prepare a hive stand, fenced against cattle or suitably screened in your garden. You may be given conflicting advice on choosing a site, but the truth is that it does not matter a great deal which direction a hive faces, whether it is in the open or under trees. The convenience of the beekeeper (and the welfare of his neighbours) is the most important consideration, subject to one or two factors. Avoid if possible a frost pocket – a patch of low ground into which cold air flows and stays on frosty nights. Also avoid exposing the hives to gales which can blow directly into the entrances. The lee of a hedge will moderate the draughts and cross winds that make flying difficult for bees coming home tired, with heavy loads of nectar for you. If any winter sun can reach the hives around midday this is good, as it will make it easier for the winter cluster to move on to other combs and also allow occasional cleansing flights at the warmest part of the day. Bees are very adaptable and will manage without these advantages if they have to, but, as my mother used to say, a little help is worth a lot of pity.

If for any reason it is not possible to start up with bees in that first spring, do not miss the opportunity of seeing bees handled by experts. Your local association will probably have a programme of apiary meetings at the branch apiary or at members' houses, with an hour and a half of beekeeping followed by tea and bee conversation for as long again. Take your own veil with you, watch, listen and ask questions. The demonstrator will be pleased to show you the queen and explain the stages of brood, pollen storage and honey ripening. You will learn how experienced beekeepers handle stocks. You may also learn by mistakes which you see other people make! After a time, ask if you may take out a frame and inspect it yourself. Learn to 'read a comb', that is to identify everything on it. Eggs are difficult to see unless the sun or brightest part of the sky is behind you; if you have reading glasses you will need them for the small print of eggs resting on the bases of cells. You will make mistakes, but we all do that – the object is not to make the same mistake twice. All the time you are gaining in practical experience, to go with all the book and lecture knowledge you acquired in the winter.

So far as your own enterprise is concerned, there is also the possibility of a partnership with an experienced beekeeper, who might be looking for a site on your land, and be prepared to share the work and the harvest from an apiary of, say, five of his hives and five of yours. Two pairs of hands make for light work and often better handling, and you will learn faster by working regularly with someone already experienced. Your colleague may have an extractor or a wax press as well as some interesting ideas to offer. If you plan some migratory beekeeping, like taking a carload or trailer full of hives to oil seed rape in April/May, or to the nearest heather in August, some help is definitely necessary.

CHOOSING A SITE

Most farms or smallholdings have odd corners of unused land where the awkward shape of adjoining fields leaves a triangle too narrow to cultivate. A fairly short stretch of pig wire across the open side with an improvised gate will make it a perfect apiary, especially if natural hedging screens one of the sides at least. A small copse or woodland area, originally not cultivated because of the trees and now left as a small nature reserve, in line with modern thinking about conservation and

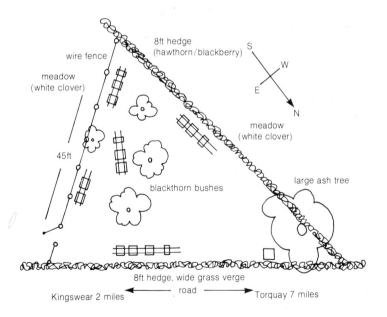

Apiary in field corner, Hillhead Farm, Brixham.

11

preservation of wild life, can also provide a good site. Some beekeepers may quote the old adage 'bees in a wood never did any good', but this certainly has not been my experience. In fact most of my honey in 1986, a very bad year, came from an apiary of nine hives amongst the trees of a small copse near the Exe estuary. After all, most of the bees in Britain lived in trees a few hundred years ago, when nearly all of the country was covered with forest. In planning an apiary site these days it is very necessary to think of the risk of vandalism, and invisibility from road or path is definitely the best security. Other humans can be more dangerous to our hives than are bears in North America.

OPENING UP A HIVE

The correct use of a smoker is one of the basic skills to acquire, and perhaps the first step is to get ready a supply of fuel. This can be a handful of wood shavings, a piece of dry, rotten wood (as found inside an old, hollow tree, for example) or just dry pine needles. Usually it is more convenient to cut or tear strips of coarse hessian from old sacks and roll them up with similar strips of corrugated cardboard, making something like a Swiss roll, which will fit loosely into the barrel of the smoker. Hessian sacks are not so common as they used to be, but are still readily obtainable from health food shops (a possible market for your honey later on) and garden centres. Get one end of the roll well alight, push it into the smoker barrel and work the bellows until smoke comes out in a steady stream from the narrow neck. After this, work the bellows occasionally to keep the smoker going, laying it on its side for a while if it smokes too much.

Wearing veil and gloves to begin with at least, direct two or three firm puffs of smoke first across and then into the hive entrance. Wait two minutes for the bees to get the message and then, standing behind the hive, lift off the roof and place it upside down on the ground or on an adjacent hive. If there is a super on, insert the hive tool under the crown board at a corner and gently lever it up, at the same time blowing a puff or two of smoke into the opening and across the tops of the frames. Lift the crown board off and lean it against the side of the hive. Make a quick assessment of the number of super frames covered by bees. Also gently lever up a central frame and lift it vertically to assess the honey content, or how well the foundation is being drawn out, as the case may be. If the frames are close-spaced it will be necessary to use the hive tool as a lever to make a small gap on either side of the frame to be lifted. Replace the frame and crown board as soon as possible, first closing up the gap. If it is necessary to inspect the brood chamber, insert the hive tool at a

Smoker plus hive tool.

corner between the super and the queen excluder and lever up gently; if the super comes away cleanly, lift it up and place it diagonally across the upturned roof. If it is stuck down with propolis then push in the small wooden wedge (to avoid crushing bees), blow in smoke and then get the hive tool into the next corner and lever up until the super can be lifted off.

Use a similar procedure to remove the queen excluder if it is rigid. If it is of the flexible zinc type, peel it off steadily from a corner. In either case look at the under-surface to check that the queen is not on it before placing it on top of the super, or leaning it against the front corner of the hive. As soon as possible, drop a cover cloth over the exposed top of the brood box, to conserve warmth and also minimise flying. Just how thorough an inspection is necessary depends on the circumstances, but assuming that a full inspection is needed, begin by rolling back the cloth to expose the first three frames, and make a small gap between the end frame and the next one by levering with the hive tool. Lever the end frame away from the side to give room, before lifting it up gently and steadily. Normally this end frame will have on it just food and only a few

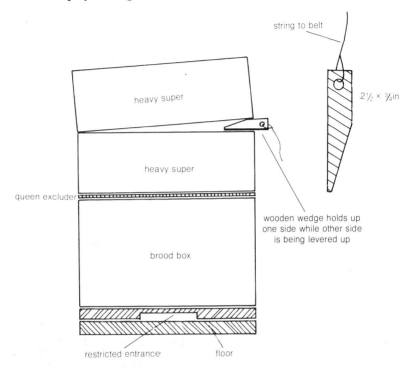

Bee safety wedges.

bees, in which case it may safely be placed on the hive stand to lean against the hive, while the second frame is levered clear of the third before lifting up. Take care not to scrape one surface against another, which would roll bees over or even crush them. After inspection this second frame may be replaced as the new end frame, to give room for the third frame to be lifted out cleanly, and so on right through if necessary. Finally, slot the first frame into the gap left on the far side and then replace the queen excluder, the super and crown board and finally the roof. The cover cloth is progressively rolled up until halfway across, then unrolled back and rolled up from the other side as far as necessary.

Skill can never be acquired only by reading, and much practical experience will be needed before all the actions described become second nature. All movements should be steady and deliberate, with no jerks or bumps, and no sudden hand movements across the top of the open hive. A working position behind the hive does not interfere with the normal flight path of the bees, so do not allow any colleague or

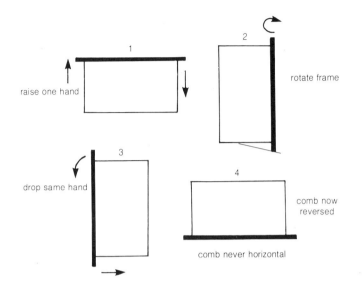

1 raise one hand

2 rotate frame

3 drop same hand

4 comb now reversed

comb never horizontal

Comb inspection.

Use of cover cloth at hive inspection.

15

spectator to stand right in front of the hive entrance. From time to time it may be necessary to blow smoke across the top of the hive, should more than the occasional bee be seen flying up. Aim to move with no delays, so that the hive is open for only the minimum time. As you will soon find out, there are some days when the bees are so good-tempered that it may be possible to take as long as you want – this will normally be on a warm day with a good honey flow. On other days, even if you do everything correctly, the bees may still resent interference, and then it is a matter for judgement as to whether you carry on or postpone the work until another day. In poor weather, it will always help to give a feed of warm sugar syrup the previous evening.

THE NUCLEUS

The above directions were given with a full colony in mind, but they will also apply to a developing nucleus. If you have just started beekeeping in the recommended way, with a four-frame nucleus headed by a young queen, proceed by siting the nucleus on the prepared stand, alongside the hive which it is to occupy. Open the entrance and leave it for twenty-four hours for the bees to get accustomed to their new surroundings. Put on a small round feeder and in the evening give one or two pints of sugar syrup (2 lb (900g) sugar: 1 pint (575ml) water). In the meantime have ready in your new brood box seven frames fitted with wax foundation, preferably choosing frames of a type similar to those in the nucleus, that is Hoffman self-spacing or metal end fitted as the case may be.

When you are ready to transfer the bees, give two or three gentle puffs of smoke into the entrance, and while the bees are getting the message, open up the new hive and make a space in the middle of the brood box. Also take out one new frame to leave plenty of room for the four frames of bees. Take the cover off the nucleus box, with a puff of smoke over the frame tops as you do so, and transfer combs plus bees to the centre of the new box, keeping them in the same order so as not to change the shape of the brood nest. Then push in the frames of foundation and shake any bees still left in the nucleus box on to the frame tops. Put the last frame of foundation at one end and replace the crown board with the round feeder over its central open slot. There may be a few bees clinging tightly to the nucleus box, so put this in front of the hive so that they can find their way in, and leave them all to settle down. Usually a few bees will be seen within minutes fanning at the entrance, heads down and tails up, to broadcast the scent with the 'come in' message.

The same evening, top up the feeder with syrup, and continue to do so every two or three days for some weeks, so long as they are taking it down. Your bees have to produce wax to build out their new combs, as well as to feed the larvae, and need all the help which you can give. Only if they have a steady supply of nectar or syrup can they develop quickly and efficiently into a full colony. For the next few days, watch the entrance, especially around midday. If all is well they will be flying freely and bringing in loads of pollen on their rear legs. If the weather is really bad, just see that they have syrup in the feeder and they will still build up.

Some colleagues may tell you to leave the new colony alone, but in my opinion it is essential for you to open up about once a week in order to learn your new craft. The young colony will normally be very docile and you will gain in knowledge and experience as they grow in strength. This is much better than learning with a full strength colony while still a beginner. In the long run, the bees will benefit by having a practised operator, and the advantage of this outweighs any minor disadvantages associated with weekly inspections. You will soon recognise the appearance of larvae of all ages, open and sealed, as well as spotting patches of eggs. These are difficult to see unless you hold the frame with light coming over your shoulder. You will see the queen on most inspections – take care that she is not on the woodwork as you replace that frame, or she may be crushed as you push the frames together.

If you hived your nucleus before the end of May, the bees should be working on eight or nine frames by the end of June, and then a queen excluder and super of shallow frames fitted with foundation could be put on. In a good year you may have three or four frames of sealed honey in the super by the end of July, but be content if they just draw out most of the combs and store some honey. You need have no qualms about taking the sealed honey for yourself, so long as the empty combs are replaced and the bees well fed with sugar syrup. It is important that the interest of a new beekeeper should be maintained, and even a few pounds of your own honey will sustain family support for the new enterprise. From just three or four frames, the bees might be brushed off with a feather and the honey extracted with a small hand-operated extractor loaned by your branch or a friendly beekeeper. All this will help to give you the experience of handling honey, and make it easier in following years when dealing with much larger quantities.

It is also possible to start beekeeping with a swarm, and detailed instructions are given in Chapter 4. The progress of a swarm hived on frames of foundation is most interesting and instructive, and most of what has been said about a developing nucleus will also apply in such cases.

2 Understanding Bees

We know that bees evolved with flowers and flowering trees many millions of years ago, in a pollination partnership where pollen was transferred from flower to flower to aid reproduction, and in return both pollen and nectar were used by the bees as food. We still have solitary bee species like the leafcutter, and bumble bees which die out every year leaving only a few young queens to survive the winter and start a new colony in spring, as do wasps. Here, however, we are concerned only with honey bees (*mellifera*) which live in communities all the year round, having a summer population rising to 50,000 or 60,000 and a winter population (at its lowest in February) of perhaps only 10,000. In India and the Far East, colonies of a slightly smaller bee (*cerana*) are still kept, and in South America and elsewhere a few colonies of a much smaller (and stingless) bee (*mellipona*) can be found. Even so, the superiority of *mellifera* is recognised everywhere and most of the world's honey is produced by these bees. In modern China two-thirds of their estimated six million colonies are now of *mellifera*, where only forty years ago the great majority would have been *cerana*. In Britain no other type but *mellifera* has ever been kept in hives, and with a population one twenty-fifth that of China we have just under a quarter of a million hives – a ratio of about one per every two hundred of our population, which is in fact much the same as in China.

Both colonies and the individual bees in them have their own separate life-span. The strength of the former ebbs and flows with the seasons but may go on for many years just as a town or village does, until perhaps it dies during winter from starvation or disease, falls a victim to predators (mice, woodpeckers, vandals) or to some natural disaster like floods, fire or falling trees. To cover such threats to their whole community, about one quarter to one third of all stocks will swarm or attempt to swarm during summer, to make new stocks to offset possible losses. To do this is as natural and as necessary for survival as the collection of pollen and nectar. In modern bee management swarming is regarded as a nuisance to be prevented, but it is important to see it as an instinctive activity and guide this the way we want to, rather than attempt just to block or prevent it completely.

The life-span of individual members of the bee colony is very varied. The queen may live for three or four years but she is usually only effective for the first two. Drones (the males), live for four to six weeks as a rule, but can live much longer if permitted – in fact a queenless colony has been known to keep a few drones over most of the winter. This is rare, and normally they are thrown out at the end of summer at the latest. Workers in summer live only about five weeks, but if reared in autumn will survive for up to six months, indeed they have to, or the colony would die out as so very few are raised in winter. It has long been known that the life-span of a worker bee resembles that of a car or plane, and depends more on work done than on age. Recent research work in Australia suggests that worker bees have a built-in potential flying mileage of about 600 before wear and tear brings their life to an end. At an average flying speed of 12–15 miles per hour this indicates about 40–50 hours of flying time.

A normal hive in summer will contain up to about 60,000 bees – these

Queen, drones and workers on mini-frame.

are mostly workers but there will also be one queen and perhaps two or three hundred drones. The queen is the colony mother, laying up to 1,500 or even 2,000 eggs per day at her peak. The drones are potential fathers, vital should the colony swarm and produce one or more young queens, each needing to mate with six to eight drones to ensure a lifetime of fertile egg-laying. The workers carry out a multitude of duties, roughly according to their age but with considerable flexibility. The following table gives a general guide but it should be understood that not all bees do all these duties in succession, and that if the need should arise, older bees can do some of the duties of younger bees and vice versa.

Duties of Workers	Age of Worker (days)
Cell cleaning; keeping brood area warm	0 – 4
Feeding larvae; attending queen	3 – 14
Fanning and ventilation control (few)	3 – 22
Polishing cells; packing pollen	4 – 24
Orientation flights, around midday	5 – 18
Secreting wax and comb building	8 – 18
Guard duties; clearing out debris	10 – 18
Collecting pollen (some bees)	12 – 26
Collecting nectar	15 – 35

(In the active season few bees live beyond five weeks.)

LIVING SPACE FOR BEES

It is important to remember that for millions of years bees lived in natural spaces in trees and caves, neither helped nor hindered by man. Only over the last 4,000 years (much less than one-tenth of one per cent of their life on earth) has man kept bees. Their basic needs have not changed: a dry, hollow space of about one and a half cubic feet, with a fairly small entrance to give reasonable shelter against cold winds and driving rain, in a position giving some protection against natural enemies (mice, wasps, woodpeckers, badgers) is all that is needed. A space twice as large as this would be occupied more successfully, with rather less swarming. A space smaller than this would give rise to more frequent swarming. Bees also need to be in an area providing enough pollen and nectar for a living. When most of Britain was covered with natural woodland and wild flowers this was no problem. Today we decide where they have to live, and some areas are very much better than others. Probably one or two hives will survive and produce a small surplus almost anywhere, but an apiary of twenty hives most certainly would not. Bees also need access to water at certain times of the year, for example in a dry spring (March and early April) when the stored honey has to be diluted before feeding to the larvae, and little nectar is yet being brought in.

FORAGE

Bees collect nectar, which is a dilute solution of various natural sugars, as produced by the flowers. Before storage this needs to be concentrated

---

Pollen and Nectar Sources

Feb.	Laurestinus, crocus, alder
Mar.	Willow*, blackthorn, celandine, flowering currant
Apr.	Laurel, gooseberry, plum*, dandelion*, cherry*, pear, rape
May	Rape*, bluebell, sycamore*, apple*, maple, field beans*, cotoneaster, hawthorne, horse chestnut
Jun.	Raspberry*, white clover*, many garden plants and shrubs
Jul.	Lime*, blackberry*, willow-herb*, white clover*, sweet chestnut, fuchsia
Aug.	Blackberry*, willow-herb*, heather*, balsam, fuchsia
Sep.	Heather*, Michaelmas daisy, golden rod
Oct.	Ivy*
Nov.	Ivy

* Indicates main sources.

Understanding Bees

The Development of a Bee (days)			
	Queen	Worker	Drone
Egg	3	3	3
Open brood	5	6	7
Sealed brood	7	12	14
Days from laying of egg to emergence of bee	15	21	24

by evaporating off most of the water content; the bees do this by fanning air through the hive. At the same time it is also transformed into simple sugars (mostly glucose and fructose) by an enzyme called invertase, produced by glands in bees.

Today the biggest single source of honey is probably oil seed rape, the crop with the bright yellow flowers which is grown to provide vegetable oil for cooking and other uses. Apart from this there are many possible sources, the more important being listed on page 21. It has to be remembered that dates of flowering vary from one part of Britain to another, being earlier in the south and south-west and a week to a fortnight later in the north and north-east. Dates may also vary from one year to another, and 1986 was notable for an exceptionally long, cold spring which caused most crops to be two or even three weeks later than usual over most of Britain.

STRAINS OF BEES

There are several strains of bees used in Britain, but in general terms they fall into two main categories: yellow bees and dark bees. Which type you choose is a matter of personal preference, somewhat like the classic choice between blondes and brunettes for partners!

Yellow or Light Bees

These usually have two yellow bands on the abdomen, but they may sometimes have three, when the bees are almost transparent in flight

22

and appear golden. They are usually good tempered and easy to handle but rather too prolific for our climate, as they tend to go on breeding when conditions are poor. If this happens, they can turn all the food on the hive into brood, so that you end up with thousands of bees and no honey. In a long, hot summer (about once in six years) they may do very well in Britain, as they normally do in Australia, New Zealand and parts of North America. The queens are large and easy to find. They are not so good at producing comb honey as their cell cappings are dull in colour.

Dark Bees

These range in colour from leathery brown to black. They are much less prolific but better housekeepers, as they regulate the breeding programme to suit the season, so performing better in a poor or average year such as we get more often in Britain. The queens are harder to find and sometimes the bees are not so good-tempered. They are much better producers of comb honey than the lighter bees. I also find that, by natural selection, stocks of yellow bees tend to get darker within a few years as successive generations of young queens tend to mate with darker drones.

CASTES

Drones

These are easily recognised by their huge eyes, larger bodies with blunt tails and their noisy flight during the warmer part of the day. They are the males of the species and have no sting. They collect no nectar or pollen, make no wax and have never been seen to clean a cell, polish the floor, fan, feed the babies or do any menial task. In modern terms they are sexist male chauvinists, who would never do any washing up, were there any in the hive. Yet no hive in summer is happy without them, and any experienced beekeeper will tell you that hives with a fair number of drones produce the most honey.

Drones are produced when the queen lays unfertilised eggs, normally in the larger cells built for that purpose by the bees some time in April, May or early June. Their essential function is to be present during the summer months so that any young queens produced may find mates, but they also help by keeping up the warmth of the hive, flying only during the warmest part of the day. Misguided attempts to trap or destroy them, or completely to eliminate drone comb, lead nowhere.

The bees just construct more drone cells, probably in the most awkward places, so that frames are gummed up with drone brood in spaces normally left free and examination of combs is made messy and difficult. It is far better to go along with the bees and accept the fact that hive morale is higher for their presence, and not to worry about the honey they consume and the food and effort apparently wasted on creatures which seem to do no work. In fact it makes good sense to provide drone comb to prevent or at least minimise the construction of these larger cells elsewhere.

Biologically, since drones arise from unfertilised eggs, they have a mother but no father; they do, of course, have grandfathers. This interesting fact means that, when in old age a queen has exhausted the stock of sperm acquired on her mating flight, she becomes an involuntary layer of nothing but drone eggs, which spells disaster for the colony. In another situation altogether a colony left queenless and without eggs or brood for several weeks may well develop a number of laying workers – here again nothing but drone eggs are laid. One of the essentials of good beekeeping is to ensure that no colony is accidentally left for long without a queen, and that old queens are replaced by young queens by the beekeeper if they fail to do this for themselves by swarming or supersedure.

The Queen

As one might expect, the queen is the most important member of the hive community. Not only does she lay over her own weight in eggs every day in summer for two or three years, but also by her glandular secretions (pheromones) she holds the colony together and stimulates her subjects to work so hard that the very name of a bee is quoted as an example to us all. If the queen is removed from a colony, the first thought of the workers is to rear other queens by feeding a few selected day-old larvae with royal jelly, that super food reserved for queens. From the half dozen or so new young queens only one survives to lead the colony, the others being killed in their cells or pushed out of the hive. Should the new queen be lost on her mating flight, hive morale goes to pieces, little foraging takes place and after a time several workers develop their vestigial ovaries and lay infertile eggs which produce undersized drones.

Worker Bees

These are female, arising from fertile eggs, yet are sexually immature because of glandular changes induced by feeding with a modified diet

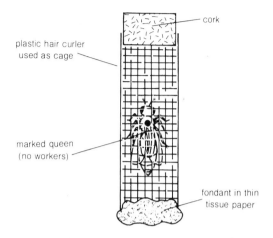

cork

plastic hair curler
used as cage

marked queen
(no workers)

fondant in thin
tissue paper

*First remove the old queen, then push the cage down
between two brood combs and leave for seven days.*

after their second day as a larva. Their abdomens contain vestigial
ovaries which sometimes develop, usually only in colonies long queenless, to produce nothing but undersized drones, fatal for the colony. In
practice this rarely happens, and good queens go on steadily laying
comb after comb of worker brood throughout spring and summer.
Their duties are varied, as shown earlier, but include the all-important
tasks of gathering nectar and pollen to feed the colony and supply their
needs during winter as well as a surplus for the beekeeper. A large
population of worker bees in summer is essential if a crop of honey is to
be obtained, yet in Britain this need is modified by the fact that in our
frequent poor summers a huge work force may consume, during a cold,
wet two or three weeks, most of the honey stored during a few good
days. Just as in racing there are 'horses for courses', so in beekeeping
there are 'bees for climates'. In more fortunate climates (Mexico or
Australia, for example) it may pay to concentrate on maintaining a huge
work force at all times, but for four years out of five our situation is
different.

Perhaps the best known fact about worker bees is that they have a
sting (unlike drones) and in defence of their home do not hesitate to use
it, though it leads to their own death. With sensible management and
protective clothing there is nothing for anyone to be afraid of, yet it is
only with experience of them that most people overcome their fear of
stings.

3 Equipment

HIVES

A typical hive has several components. From the bottom up these are: the floor, with an entrance block regulating the size of the opening into which the bees fly; the brood box or boxes; a queen excluder or grid, which allows the workers through but keeps the queen and drones below; one or more shallow boxes called supers to hold the frames in which the honey is stored; a crown board and finally a ventilated roof. A hive stand is needed, which can be just four bricks or an old milk crate, although it is much better to use two metal runners resting on two concrete blocks, giving room for a pair of hives about a foot apart. A wooden hive must never be placed directly on the ground or the base will be damp, which is bad for the bees and also leads in the end to rotting woodwork.

It must always be remembered that the important dimensions of a hive are internal; the actual outward size depends on the thickness of the wood used. The internal dimensions are so important because the frames have to hang from rebates in the side walls with a bee space of about a quarter of an inch between frame and wall, also between frames in boxes placed above and below. It is normally assumed that hives are made of wood, but other materials have been used, such as fibreglass, hard plastic, and even concrete. Wood still has many advantages, however, including the fact that it lessens condensation and allows a hive to 'breathe' to some extent, which other materials do not. A final point is that some hives (for example, British National) are not normally made with a top bee space, that is, the frame tops are flush with the top of the box leaving a quarter of an inch clearance below. Most other hives (Langstroth, Dadant, British Commercial) are made with a top bee space, that is, with a quarter of an inch space at the top of the box, and the frames flush at the bottom. There are advantages and disadvantages in both systems, but for obvious reasons all the boxes on any one hive must be compatible, otherwise there will be a double space or no space at all between boxes for the bees to circulate.

Hive components.

Equipment

Choice of Hive

There is a bewildering range of various types of hive and bees will live happily in any of them, so the main consideration is really the convenience of the beekeeper, although a number of points should be remembered.

Initial Expense

This will depend on the complexity of construction as well as on the size and nature of materials used. In Britain it has become customary to use western red cedar, the lightest and best but one of the most costly woods on the market, whereas in most other countries cheaper timber is used, protected by emulsion paint, Cuprinol or some such preservative. Some hives are double-walled, like the WBC, with a clinker-built outer casing usually painted white, making them much more expensive than a single-walled hive, but better looking.

Availability Secondhand

A type of hive in wide general use is more easily bought secondhand, and also more easily sold later on if one decides to do so. In Britain the commonest hive is the British National, followed by the WBC (both using the same size of frame, the British Standard). Since the WBC used to be more popular years ago when prices were very low, they are often available secondhand.

Simplicity of Construction

If you have a minimum of woodworking skill it is possible to make your own hives – simple types like the British Commercial and the Scottish Smith Hive are much the easiest to make.

Migratory Use

Some hives are more difficult to transport than others, if you intend to move hives to oil seed rape or heather for an extra honey crop, or to fruit orchards for a pollination fee. Any double-walled hive (WBC or Burgess Perfection) is much more awkward to rope up and move. The larger and heavier hives like the Dadant are more trouble than the lighter Nationals. Smith and Langstroth hives with their rectangular shape are easier to fit into the confined space of the average hatchback car.

Appearance

If you wish to have just one hive on the front lawn, then a white-painted WBC or even a straw skep as used last century (and still most commonly depicted on honey labels) are both more attractive than square or rectangular boxes, and passers by will say, 'That looks like a real beehive!'

Standardisation

It is wise to choose one type and then to stick to it. Nothing is more exasperating than to have different types of hive in the same apiary, and find that frames, honey supers and roofs are not interchangeable. It is easy to give this advice, but in real life you may sometimes have the unexpected opportunity of a real bargain, say three hives of lovely bees but in a different type of hive. The answer is probably to go for British Nationals as they use the same frames as the older WBCs, and these two types together account for 80 per cent of the secondhand market, so that bees on their frames are easily interchangeable.

My own first choice would be for either British National or British Commercial hives. Although the latter have larger frames, the outer dimensions of component boxes are the same, to within a quarter of an inch, so that supers, queen excluders, crown boards, floors, and roofs are all compatible.

Approximate Comb Areas of Various Frames	
Frame	Area
British Standard Deep (also WBC and Smith)	5,000 worker cells
British Standard Shallow (also WBC and Smith)	3,300 worker cells
British Commercial Deep	7,000 worker cells
British Commercial Shallow	4,500 worker cells
Langstroth Deep	7,000 worker cells
Langstroth Shallow	4,500 worker cells
Dadant Deep	8,500 worker cells
Dadant Shallow	4,300 worker cells

Equipment

Internal Volume (Single Brood Box)	
Box	Volume
National	1.25 ft³ (0.035 m³)
Langstroth	1.48 ft³ (0.042 m³)
Commercial	1.60 ft³ (0.045 m³)
Dadant	2.10 ft³ (0.059 m³)

General Points

Most hives take eleven frames in each box, but WBC hives take only ten. Over the world as a whole, the most commonly used hive is the Langstroth; in fact in the USA, China, Mexico, Australia and New Zealand they just speak of a hive, and take it for granted that it is of this pattern. No mention has so far been made of the Catenary Hive, having a curved brood box following the shape of a comb built naturally by the bees; or the Long Hive where the bees expand sideways instead of upwards; or the Double Seven Hive which has all boxes and frames of the same size, for both honey and brood, the size of each component being halfway between that of a National Deep and Shallow. These and other interesting variants can be tried by enthusiasts but my advice to farmers and smallholders remains – go National!

It is usually best to treat exterior surfaces with a good preservative, like Cuprinol or creosote, avoiding any containing insecticide, and taking care to air well for a week or more. Normal paint stops the wood 'breathing' and can lead to moisture lifting the paint in bubbles. Emulsion paint is favoured by many beekeepers, but mostly the professionals use creosote or Cuprinol. A double-walled hive like the WBC can be painted with brilliant white gloss paint with no problems, as the inner boxes are left untreated.

Queen Excluders

In wild bee colonies, stored honey is usually found above or on the outer combs, but tends to be mixed up with brood cells, pollen and eggs – nutritious though the mixture may be it would not appeal to many people nowadays. Queen excluders are designed to allow worker bees free access to the honey supers above while confining the queen (and drones) to the brood box below. This separates the nursery and colony store-room below from the owner's pantry above. There are two main types in use in Britain today, the slotted zinc (occasionally found made

Queen excluders, showing different types of grid.

in strong plastic) and the Waldron framed wire excluder. A German variety (the Hertzog) is much stronger than the Waldron and is gaining in popularity. All have slots or gaps of 0.165 inch (4.2mm). We owe the original invention to the Abbé Collin (France 1849).

In countries like New Zealand, with larger and more consistent honey flows, commercial beekeepers often work without excluders, and the pressure of honey stored above pushes the queen down to the brood area. Unfortunately our honey flows in Britain are much less dependable, and five years out of six a queen given access to all boxes would 'chimney' up the middle and lay her eggs all over the place, so that grubs and eggs would be found mixed up with honey in the supers.

31

Equipment

Flexible Slotted Zinc Excluder

Most amateurs use this type, which rests directly on the tops of the brood frames, so that only about a quarter of its area is available to the bees. It gets stuck down with propolis (bee gum) and can only be peeled off with difficulty, disturbing the bees unnecessarily. Also in scraping the surface clean it is easy to catch up a piece of metal and distort the opening, leaving a space through which a slim queen can squeeze. Another practical difficulty is that the bees tend to force propolis between the zinc and the frame tops, pushing up the flexible zinc so that the bases of the frames above cannot get down to their proper depth. The frame lugs then do not rest correctly on the rebated tops of the supers, and lean over sideways, resulting in distorted combs which are awkward to handle and are often stuck to each other. Should this type of excluder be used on a hive with a top bee space, then it sags in the middle and gets stuck down over a wide area with burr comb and propolis. Not only framing but also reinforcing slats at 6 inch (15 cm) intervals are necessary to make an ordinary zinc excluder effective. Short slot excluders are slightly stronger than long slot ones, but both should have been phased out years ago. A much more rigid steel slotted excluder is now on the market and this is a great improvement.

Waldron Excluder

This type has been on the market for many years, but for some inexplicable reason is not framed with a bee space on one side only, but with half a bee space on both sides. This type of excluder normally collects a good deal of burr comb, and also cell comb extending down from frames, but at least it is strongly made so that it can be levered off without distortion.

CONSTRUCTING A HIVE

Wild colonies have often been found in quite small spaces, with access to another space via a hole of as small as one square inch ($6\frac{1}{4}$cm^2), and this apparent barrier has not hindered the bees from building ample combs beyond it. Based on this principle I have made cheap and efficient excluders by framing a sheet of plywood so as to give a bee space of a quarter of an inch on one side only, and then inserting a strip of zinc or wire excluder across the centre, as shown on page 31. Such excluders should be put on hives so that the slots or wires are at right angles to the brood frames below, in order to give equal access to the supers to bees from all frames. Experience over many years has confirmed the success of this type, and the ease with which full supers may be separated from these excluders when taking off the harvest is in sharp contrast to the difficulty of getting heavy supers off zinc excluders, which often stick and have to be levered down, distressing and annoying the bees, who take their revenge. One old and damaged excluder, zinc or wire, may be cut up into five or six strips to make that number of excellent excluders, saving money and getting a more efficient product.

Roofs

A good roof needs to be solidly made with a metal top and effective ventilation. It helps to have a two-inch deep fillet around the inside to lift it up from the crown board and also provide an air space. Drill a couple of one-inch holes, gauze covered, on opposite sides, leading to dormer roofing bulges beaten out of the metal roof before fixing it to the hive. There will then be real ventilation, with no condensation. Orthodox roofs have tiny slots covered by the metal part so that in practice they are almost completely ineffective, and are soon blocked by spiders and debris anyway. A roof such as that illustrated cannot be bought and will have to be home-made. If you are making one, then it is worth while to put a piece of quarter-inch thick polystyrene centrally between metal and woodwork, over about half of the area, to give valuable insulation and also a slight camber to the roof-top, so that rain drains off more readily and bees do not get stuck down by their wings on a wet surface. If you have to manage with a normal roof you may get some condensation but bees have a knack of surviving whatever we do, or fail to do! They will come through the winter more strongly with a roof such as I have described.

Entrance Blocks

All hives are supplied with entrance blocks of one kind or another, to give the choice of a full width or restricted opening. That shown in the drawing is my own design, which also doubles as a mouse guard, and being hinged to the floor board by a long screw is never lost or mislaid, any more than your front door is. Made longer to extend to the side of the hive, it is prevented from being pushed in too far. It also enables great leverage to be exerted with a hive tool to open it when tightly stuck in wet weather. The palisade of nails at 9mm centres totally prevents the entry of mice in winter, yet provides ample room for the bees to get in and out with no impediment whatever. In spring the orthodox mouse guards, with their circular openings, tend to scrape

Equipment

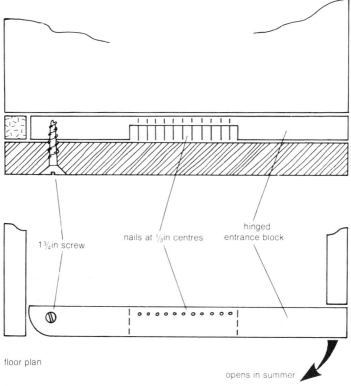

front view

1¾in screw

nails at ⅓in centres

hinged entrance block

floor plan

opens in summer

Mouse-proof hive entrance.

some of the pollen from workers' legs, at a time when the hive most needs all the pollen it can get. This does not happen with the nails, as they are vertical.

Frames

There are several different types of frames on sale. My own preference is for Hoffman self-spacing frames, both deep and shallow, with wide top bars (wedge type) and grooved side bars. The type of frame which has separate metal or plastic ends should be avoided at all costs. These metal ends get stuck down with propolis, easily distorting with pressure, cut the fingers when pulling them off and are generally a nuisance. If non-Hoffman frames have been acquired, buy some Hoffman converters, made in strong white plastic, which can be clipped to the side bars. The deep brood frames will always have to be close-spaced, and with Hoffmans they may be pushed along together without crushing bees in manipulation. The shallow frames have to be close-spaced when carrying the original sheets of wax foundation, otherwise wild comb will often be built in between. Once drawn out into combs it is good practice to reduce from eleven to nine frames, either spacing by estimation, or using castellated metal runners with nine slots, or tacking small studs on the side bars. This results in fatter combs and less work for the same amount of honey at extracting time, plus the bonus of spare drawn frames which can be used to make up an extra honey super with no extra expense.

Wax Foundation

It is possible to make this yourself if you have the wax, using either a wax press or the technique of first making thin, plain wax sheets and

Fitting frames with wax foundation.

then imprinting the hexagon cell shapes by passing them, sandwiched between a pair of plastic formers, through the rubber rollers of an old laundry wringer. However, most people still buy sheets of ready-wired foundation of the size needed and fit these into wooden frames as required. Wired foundation is essential in both brood frames and honey supers, unless you are aiming at cut comb production, which is described later.

HONEY-HANDLING EQUIPMENT

With just a couple of hives it makes sense to borrow or hire an extractor from your local beekeepers' club while waiting for an opportunity to buy a good secondhand one at a bargain price, or gaining experience on various available models before spending a large sum on a power-driven nine-frame radial. I managed for many years with a hand-operated six-frame tangential extractor bought secondhand for a few pounds, but sadly you would have to pay much more now. You will also need an uncapping knife and fork as sold by all the bee appliance dealers, although the former can be replaced by a thin, sharp, wavy-edged kitchen knife. A purpose-made honey tank with built in filter top and a special honey-gate tap is almost essential for holding and bottling the crop, but to begin with you can manage with large plastic buckets (from your local baker) and jugs. If you have more than four hives you will almost certainly store the honey in 28lb tins or buckets and bottle just your immediate requirements. Honey stored like this will set quite hard after a time and gentle heat will be necessary to liquefy it before bottling. The photograph and instructions on page 99 show how to make a honey-warming cabinet made from an old fridge fitted with a 60 or 100 watt bulb as a heat source.

PROTECTIVE CLOTHING AND OTHER ITEMS

When you have firmly made up your mind to go in for bees, it is strongly recommended that you buy a modern white bee hood built into a jacket to wear with trousers tucked into wellies for one hundred per cent protection. Together with a pair of soft leather or rubber bee gloves, this will give you self-confidence in the early stages of bee handling. Later on a light bee veil tucked into your normal clothing may well be enough for casual work with or near hives. A hive tool is also essential, and there are two types in general use. If you mislay it, a broad screwdriver will serve the purpose but is much more likely to damage

Radial honey extractor and bottling tank (stainless steel).

the woodwork, and is difficult to use for scraping off wax and propolis.

A hive smoker is essential, and I would recommend one with a copper or stainless steel body, to give long service with no rust. Choose a large one which will not need constant refuelling when working on a number of hives. For your workshop buy a blowlamp for scorching floors and bee boxes acquired secondhand, or after taking out of use in your own hives, plus a wire brush and a stiff-bristled brush for cleaning up generally. One small tool, a rampin for pushing in small nails, makes frame assembly and waxing much easier.

Equipment

Hive Stands

Strongly recommended is the type shown on page 27, comprised of concrete blocks with iron or stout timber rails. As it may stay in position for many years it is well worth the trouble of levelling (using a spirit level) and leaving a gap of at least 6 inches (15 cm) below the rails so that you can get a rake or hoe to leave a clear air gap below, avoiding damp floors.

Carpet Squares

These are very easy to cut out from old carpets which would otherwise be thrown out, and 18 inch (45 cm) squares have a variety of uses: as temporary covers when inspecting a hive and taking off supers; as top covers over crown boards when roping up hives to move short distances (up to two hours' drive – except in hot weather, when ventilated gauze covers are necessary), and also as hive insulation in March and early April. The tough canvas/hessian under-surface resists chewing by the bees and the softer pile on the upper surface provides insulation. Glass wool attic insulation and white polystyrene sheets cannot be placed directly over crown boards as the bees will chew away particles from both these substances.

Small Wooden Wedges

Preferably attached to lengths of string, these can be clipped to your belt and are most useful for holding open a gap when taking off supers. Once a hive tool has made an opening, some bees crawl into the new space and are crushed if the supers are allowed to go down while the beekeeper is breaking the propolis seal at another corner.

Cover Cloths

These are important when inspecting hives at any time, but most of all in spring when it is important not to chill the brood. They should be of soft material, about $18\frac{1}{2} \times 21$ inches with a hem at each end to hold slats of wood to keep the material stretched flat, and also to hold the cloth down in a breeze. The rigid slat also makes it easy to roll and unroll. Ideally a pair of cloths enable a narrow space to be uncovered at any time, but in practice you can manage reasonably well with just one (*see* page 15).

Swarm catching equipment.

Swarm Catching

For swarm catching, when the bees have clustered, you will need either the traditional straw skep or a strong cardboard box of about eight gallons (20–35 litres) capacity, with no loose flaps inside. You will also need a square piece of hessian or some strong material of open weave (a square metre or larger), a small bow saw, a pair of secateurs, and a bee brush (or large, strong feather). A hair curler with a cork to fit it will hold the queen if seen. Finally, make sure you have a length of thin rope, an extending ladder on the car roof rack and a frame of old, black brood comb.

4 Spring Management

During the four months of November, December, January and February usually regarded as winter, the bees will have been semi-dormant in a cluster, slowly eating their way through stores of honey or ripened sugar syrup. On mild days some may have been flying briefly around midday even in January, but the purpose will have been cleansing flights to evacuate bowels rather than store gathering. In favoured districts in some years there may have been a small amount of pollen coming in (from late ivy in November; from laurestinus, helleborus or very early prunus in December to February) but almost no nectar. However, the queen will usually be laying again on a small scale by January and by mid-February brood rearing will be increasing and food stores will be consumed much more rapidly.

Even when the air temperature is as low as 42°F (6°C) some bees will be fetching water to thin down stored honey to feed the larvae. Most years it rains enough to leave drops of water on every blade of grass or leaf, so the bees do not have far to go. The difficulty arises in a cold, dry spring such as we had in 1986. Thus one of the first jobs of spring is to ensure that a source of water is available in a sunny corner within 25 yards of the hives from mid-February onwards. The bees have to carry water in their honey stomachs, so have no fuel other than the sugar content of their body fluid. This means that they can only fly very short distances, so the water must be close and preferably not ice cold. In practice an old trough or bowl of peat, standing where the sun can reach it, is ideal. In theory it might appear sensible to put on a feeder with dilute sugar syrup at this time, but in fact this is found to excite the bees, which fly out in large numbers looking for the source of nectar, as it seems to them; some get chilled and fail to get back home, at a time when every bee is needed. This is why sugar syrup should not yet be fed when a hive feels light and possibly short of food. Candy is much better at this time, and in recent years I have used bakers' fondant instead of making the candy myself.

The fondant is manufactured from finely ground white sugar plus a little water and glycerine, and is eagerly taken by the bees. Almost any baker will supply this, in 12½kg packs at a price not very much more expensive than ordinary packet sugar. The fondant cuts easily and can be pressed into a plastic container (the sort which hold ice-cream or

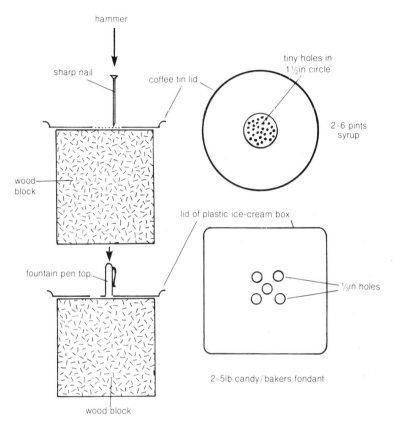

Home-made feeders – lid detail.

margarine) and inverted over the feed hole of the crown board. Without a lid the fondant may slowly flow down between the frames, so it is best to punch four or five holes approximately 1cm in diameter in the lid before inverting it over the feed hole. I make the holes by punching with an old fountain-pen top, the lid being placed on a wooden block, as illustrated. The lids are fortunately shaped to give a depression of about a quarter of an inch, which is perfect as a bee space. Experience over several winters has shown that these holes give just the right area of contact for the bees, which will happily eat their way into the fondant and there is no risk of it sliding down to make a sticky nuisance on frames below.

It is certainly not necessary to feed every hive, just those which feel lighter than the rest, as an insurance against starvation until it is safe to feed sugar syrup again, usually in the second half of March.

During winter there is a net outflow of water vapour from the hive, as honey and stored syrup are slowly used up, the end products being carbon dioxide and water vapour. During the second half of February the water balance changes as more and more eggs are laid and the need for larval food increases. Any condensed water is recycled and also some water has to be brought in. By the end of March this process increases still more. Whereas in winter the hive needs slow through ventilation to prevent condensation, by March this is no longer so important, and any condensed water will be used up. What is important in March and April is to provide some insulation, at a time when a limited number of bees are finding it difficult to incubate a much larger area of brood, and the cluster is over-stretched. A hand placed on the crown-board will detect the heat which is being lost. The best help is a square of glass wool insulation, or white polystyrene foam, preferably over an 18 inch (45cm) square cut from a tough old carpet to prevent the bees from teasing away at the main insulation. If it is necessary to feed, then a hole can be cut in the insulation to accommodate the feeder, or strips can be laid around the feeder when it is in position.

POLLEN SHORTAGE

In a normal spring, bees will be bringing in pollen from laurestinus, crocus, celandine, early prunus, blackthorn and above all from the pussy willow catkins, which also yield nectar. Pollen from hazel is only of minor importance and bees are not usually attracted to it in any numbers. In a year with really cold weather in February and March (like 1986, for example), the limiting factor is the shortage of pollen to provide protein for the growth of the larvae, and in such a year hives fail to build up as they normally do.

In my own case (1986) hives taken to Dartmoor for the heather the previous August/September had brood and bees across seven frames by the beginning of April, while no other hives had half this quantity of brood. Heather honey is unique in that it contains about two per cent of protein in the actual honey, quite apart from any stored pollen, and on the moors bees are almost always able to fill their brood frames, even when there is little or no surplus in the supers for the beekeeper. So far as they can, bees do store pollen in summer and autumn for their own use in spring, and winter bees have protein and fat stored in their own bodies as well, but often there is not nearly enough. Without access to either fresh pollen in spring, or soya flour substitute, the normal build-up is slowed down considerably.

SPRING FEEDING

Beekeepers differ about the wisdom of spring feeding, and many will say that so long as there is plenty of food on the hive, it makes no difference whether you feed or not. In some years this may be true, but my own view is that it helps almost any hive to be given a gallon of sugar syrup about the middle of March, for two reasons. Firstly, in a cold, dry spell this also provides all the water they need, as even thick syrup contains nearly 40 per cent water. Secondly, bees, unlike some humans, seem to understand the difference between capital and income, and build up more readily on a steady income of fresh nectar or syrup than on stored food. In an exceptional March I have known such an income from pussy willow that colonies have not only built up in record time but have also produced a super of capped honey which could be taken off and extracted in April. I should add that this has only happened to me once in the last 25 years.

OIL SEED RAPE

In the last dozen years this crop has been increasing so fast that as you drive across much of Britain in late spring or early summer, fields of bright yellow flowers are now seen everywhere. It has the advantage, from a farming point of view, of being a good break crop as well as being very profitable to grow – it is subsidised by the EEC to ensure that we are self-supporting in vegetable oil for cooking and for several industrial processes. There is no doubt that this crop is now the biggest single honey producer in Britain. Nectar from the rape has a high sugar concentration, from thirty to fifty per cent, with a generous supply of pollen as well. Bees love it and will fly up to two and a half miles to collect its nectar. Estimates of possible honey yields vary from 90 – 200 lb per acre (100 – 230 kg per hectare), and two full supers per hive (60 lb honey) is a realistic expectation from well managed stocks kept at the rape for a month or five weeks.

With modern varieties of rape flowering before the end of April, it is necessary to have colonies stronger earlier in the year than normally happens naturally. The main limiting factor in most areas is the shortage of protein needed for building up the food glands of young nurse bees, and hence the tissue of larvae. Roughly speaking, to provide enough protein for its body each worker needs about its own weight of pollen, so that a pound of pollen will produce about 4,500 bees. With young queens mated the previous summer and hives taken to the heather, or in areas with a mild winter and an abundance of early spring flowers, some stocks may be strong enough to take advantage of the rape without special preparation, but over most of Britain it is necessary to feed a pollen substitute as well as a gallon of syrup in March.

A suitable recipe is 3kg of soya flour to 1kg of brewers' yeast, soaked in 8 litres of thick sugar syrup and thoroughly stirred before leaving overnight. The consistency of the dough-like candy obtained should be such that it will stay on the frames without running down. Wrap each individual flat patty of about $\frac{1}{2}$kg in cling-film and puncture in several places underneath before placing on the brood frames over the bee cluster. As soya meal is too coarse for the bees, only de-fatted flour should be used. The addition of a small amount of natural pollen (trapped the previous summer and stored in a deep freeze) greatly improves the value and attractiveness of the mixture to the bees.

At the rape normal practice would be to set up stocks initially with two supers over the queen excluder and a single brood box. As excellent combs will be drawn on the rape flow, one box could with advantage be of frames fitted with wax foundation. After the first fortnight, be ready to put on extra supers as necessary. Close co-operation with farmers concerned is necessary to avoid disaster through spray damage, and the danger period is at the end of flowering time, as contractors will often press on with spraying when there are still five to ten per cent of yellow flowers left. It probably pays to move away while the bees are still able to work with 20 per cent of blossom, to be on the safe side. This may also make extracting easier, as explained below.

Extraction of Rape Honey

Although extraction is normally done at the end of the summer, when honey has been well ripened on the hive, rape honey is an exception, as it granulates rapidly while still in the combs and has to be extracted immediately the supers are full. This is because it contains more glucose and less fructose than other honies. My own experience is that the supers may be left on while the bees are working 20 per cent or more of blossom, but if by chance they have swarmed and are no longer in the top supers to keep up the temperature, there is a risk that the honey will granulate. Normal clearer boards take forty eight hours to be effective, which gives the honey time to granulate, so a special technique is necessary to be certain of success.

Use Canadian-type clearer boards, as shown in the diagram. These are easily made at home and are effective in a few hours – in fact they must not be left on longer than this or the bees will learn how to get back. Remove the supers within five or six hours and immediately stack them over a honey warmer to keep the temperature up to at least

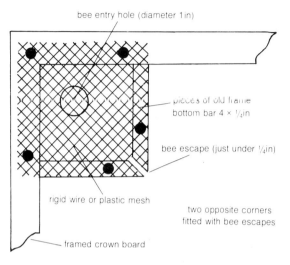

bee entry hole (diameter 1 in)

pieces of old frame
bottom bar 4 × ¼in

bee escape (just under ¼in)

rigid wire or plastic mesh

two opposite corners
fitted with bee escapes

framed crown board

Canadian clearer board - rapid.

75°F (24°C). Extract within twenty four hours and either bottle at once, while still quite warm, or run into 28 lb tins or 30 lb bakers' pails, which can be stored and warmed up in a cabinet whenever needed for bottling or blending. It can be quite disastrous to leave the honey in the extractor overnight, as granulated honey takes many hours of tiresome work to clear. Should honey granulate in the combs, there is then no way of liquefying it without destroying the combs – the choice is between laboriously scraping down to the mid-rib or cutting out the whole combs and melting them down so that the liquid wax lies on the top of a layer of liquid honey, from which it can then be separated. It is most important to tap off the hot liquid honey and cool it as soon as possible; also do not exceed the temperature of melting wax (145°F or 63°C) as honey is seriously damaged within a few hours at this temperature.

My personal preference is to blend equal parts of rape honey with last year's medium or dark honey. This gives a lovely golden liquid, with a good bouquet and deep flavour, slow to granulate and very popular with customers. Rape honey by itself is very sweet but rather lacking in character, granulating rapidly to a solid white mass resembling lard. However, some customers take it and even ask for it, especially if it has been creamed, by a technique to be described later (*see* page 100). If you have just an occasional super of hard-set granulated honey and wish to retain the combs, one alternative is to place it *underneath* the brood box

45

of a strong stock for the rest of the summer, when the bees will gradually empty the combs and use it themselves, or blend it with the normal summer floral honey to be taken off the top boxes at the end of July in the usual way. Usually the rape honey super will be empty before the end of the main summer honey flow, as bees do not seem to like honey stored below the brood nest, and the empty super can then be moved up to the top of the hive to collect the surplus, if any, arriving in the second half of July or in August on the heather.

POLLEN TRAPPING

At the rape there is ample pollen for current bee needs and it is good practice to fit a pollen trap between the floor and the brood box for

Pollen trap, with pollen in tray.

Spring Management

two weeks. The photograph shows one with the sliding drawer containing about a pound of pollen pellets. Although some books speak of the need to dry pollen, I just deep freeze it immediately, and either use it the following year with soya flour in spring as already described, or use it for making up my special product, natural pollen preserved in honey, which is in great demand (*see* page 88).

MOVING HIVES

If there is no major crop of oil seed rape within two miles, it may be well worth while to move even just two hives to a temporary site close to the field, by friendly arrangement with a local farmer. Since rape is self-pollinated by wind, and insect pollination is not vital, do not expect a fee such as you would get if moving to a fruit orchard. However, recent experience has shown that rape flowers visited by bees produce a more uniform size of seed, in a slightly shorter time, with more even ripening and easier harvesting, though with only a very small increase in the total weight of the crop.

There is not much danger of hives overheating in April, so moving hives to the rape is much simpler than moving to the heather at the end of July or the first week in August. If the journey is expected to take not more than an hour I would suggest just blocking the entrance with a strip of foam rubber, tapping in a few hive staples to link the boxes and roping up as with a parcel. For a longer journey on a hot day, fit a gauze screen board (with no carpet square). I have found 18 inch (45 cm) squares of old carpeting very useful, as one of these placed over the crown board provides some warmth as well as preventing the escape of bees. In roping up I use a length of thin rope with a loop woven into one end, through which the other end is threaded and pulled tight with a knee pressed against the hive. I then secure it with a couple of half-hitches, passing the rope around the hive again at right angles, pulling it tight after looping under the first length and again securing. A temporary stand is needed at the rape, such as an old pallet, two pieces of thick wood, an old tyre or a milk crate, just to keep the hive off the ground. Do *not* place the hives into the actual field or you may have difficulty even in finding them on your next visit. The edge of the field or even an odd corner of land a short distance away will be just as easy for the bees.

Roping a hive for transit.

Removal of Hives

As previously mentioned, it may well be worth while to bring hives back while the crop is still almost 20 per cent in blossom, before the possible spraying and also to lessen the risk of swarming. Last year I left mine on a day or two too long and had to face an emergency move late at night after a last minute warning of a spraying programme due the next morning. Fortunately I had already taken off the honey and had planned to get the hives back the next day. One of my hives had also swarmed, but having placed a bait hive 75 yards away in a hedge on a downward slope, I was lucky enough to find them in it and brought them home with the main hives.

The weight of honey is such that it is difficult to move hives plus full supers. I recommend clearing the supers down into a fresh empty super

Spring Management

placed immediately over a queen excluder, as there will not be room for all the bees in the single brood box. On return from the rape, a strong colony may well be preparing to swarm, and I believe in applying an artificial system of swarm control at this time (*see* page 69).

THE FIRST SUPER

It is most important to have the first super on, over a queen excluder, by the middle of April. Some books suggest that you let the bees tell you when to super up, and say that the presence of white, new wax at the top of the brood chamber or first super is the indication that more room should now be given. I would say that when you see this, the bees are telling you that you should have had that super on ten days ago!

In beekeeping the time is always later than you think. After weeks and weeks of cold English spring and very little apparent development, a sudden spell of warm weather brings out all the dandelions and spring flowers, and within days bee activity is trebled and the hives are clogged with thin nectar spread out over many combs before being evaporated to a third that quantity of mature, ripe honey. Get that first super on before the middle of April, and if you are going to be busy or away from home, put on two, with a sheet of newspaper between them. This will prevent loss of heat by convection, and when the bees need extra space they will soon chew through the paper and occupy it. The first super should always be of drawn comb, so that storage space is available in good time for any sudden flush of nectar. The second super can be of wax foundation. Quite apart from nectar storage space, in a poor year there is a need to provide room for the bees themselves, to prevent overcrowding, which can lead to unnecessary swarming.

5 Summer Management

In the period from May to July the main activities of beekeeping take place – this is when most of the honey comes in, when the bees may swarm, when nuclei can be made and queens reared. A farmer is also very busy with work of one kind or another and so must manage his bees with a minimum of time. Perhaps the single most important thing to understand about bees is how and why they swarm, and what we can do about it.

SWARMS AND SWARMING

The reproductive cycle of bees might at first appear to consist of egg-laying by the queen, brood rearing by the workers and building up of the hive population from about 10,000 in mid-February to 50,000 or more in mid-summer. This is indeed reproduction on a grand scale, but by itself it merely ensures the growth of surviving colonies. Every year a small percentage of colonies dies out – from starvation, forest fires, hives being knocked over by animals, mice getting in and eating their food, disease and nowadays from vandalism. Before bees were kept in hives they lived mostly in trees, with a minority living in caves. They existed long before humans, but even then some colonies would be destroyed every year – by bears, falling trees and so on. Without swarms to increase the number of colonies, bees would all have died out millions of years ago. It is important for all beekeepers to understand that the swarming impulse is programmed into bees for their own survival as a species, and although some races may swarm more than others, do not believe anyone who offers to sell you a hive of 'non-swarming bees' – there is no such thing!

Although swarming is so basic, it does not fit the convenience of modern beekeeping practice, and we are able to breed queens and make increase in stocks without waiting for swarms. There are many techniques for so-called 'swarm prevention', but usually these involve cutting out queen cells at a time when the queen has already scaled down her egg production, so that the potential work force is going to be decreased at a time when we need the maximum number. Or perhaps the control method involves taking the brood nest to pieces every few days in summer to check if there are any swarm cells, with disturbance to the bees and consequent loss of honey production, decrease in egg-laying by the queen and so on. It is better to *go with the bees* and help them to do what they want to do, but first we need to analyse the situation.

What is a Swarm?

When a colony first swarms, about one third to one half of all the bees come out and fly round and round in a cloud until joined by the queen. They then settle with her in a large cluster, usually only 20-50 yards away, probably quite close to the ground, maybe on a fence, fruit bush or low branch of a tree. Bees only come out as a swarm when at least one of several queen cells in their hive has been sealed; bad weather may sometimes delay a swarm several days, by which time more than one cell may not only be sealed but ready for the virgin queen inside to emerge.

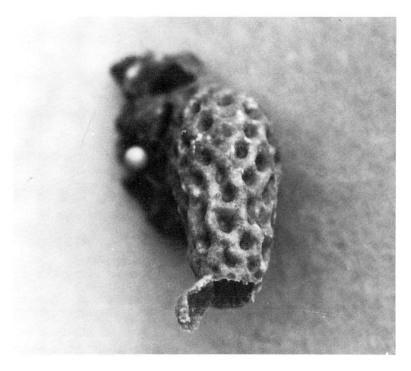

Queen cell, showing natural emergence.

Summer Management

If the swarm has not yet left, the workers will deposit new wax to seal the young queens in as fast as they try to bite their way out, but then in the excitement of swarming this is forgotten, and sometimes one or two virgin queens may accompany the swarm with the old queen.

The prime swarm with the old queen may hang in a cluster for only an hour or two, or sometimes for two or three days, before flying quite a distance and clustering again, or else being lead by scout bees to a site chosen as their new home, whether in a hollow tree, roof space of a house or empty hive. A few days later a second swarm, called a *cast* may emerge, with one or more virgin queens. The cast is usually rather smaller and often flies further away, and settles higher, than the prime swarm. Sometimes a second cast may come out a few days after the first, before the original colony settles down to build up again under a newly-mated young queen.

This procedure is really colony reproduction, necessary in nature for

A fine swarm in June, home apiary.

the survival of the species, but upsetting for the beekeeper, who would prefer to keep his work force together in order to get a good honey yield. Although strong stocks in southern counties have occasionally swarmed in the last week of April, this is unusual. In most years there will be some swarms in late May, especially in stocks brought back from the rape, but most swarms will arise in June and the first week of July.

Swarming Factors

Age of Queen

An old queen is much more likely to produce a swarm than a queen in her first full year of work, and for this reason most professional beekeepers all over the world re-queen every two years. We used to say years ago that if a colony was not satisfied with a queen's performance, they were more likely to swarm. Dr C. Butler's work at Rothamsted in the 1960s showed the importance of what he called *queen substance*, a pheromone produced by the mandibular glands of the queen and distributed throughout the colony from bee to bee by food exchange. Amongst other things, bees getting a normal ration of this substance do not build queen cells, but older queens secrete less of this pheromone.

Overcrowding

This is another factor which may induce swarming, not only because there are more bees to share a limited amount of queen substance, but also because it is more difficult to distribute it when every space in the hive is crowded.

Bad Weather

A period of poor weather, in which bees cannot fly, at a time just after a few good days with much nectar available, can also help to induce swarming. Again this is possibly because the hive is overcrowded, with many bees fanning, or blocking the spaces between the combs as they sit with drops of nectar on extended tongues to evaporate it down.

Lack of Egg-Laying Space

This can be a contributory factor, as, if supers are not put on soon enough, the bees have nowhere to dump a sudden flush of nectar, and so have to store it in empty brood cells which the queen needs for egg-laying. Then the queen has to reduce her egg-laying so is fed less, which also results in a reduced output of queen substance, leading to queen cell production.

A sensible approach to management is to re-queen regularly, make sure that the brood chamber has combs of good quality and is not clogged with too much unused winter food, put on supers well ahead of need rather than too late, and then accept the fact that an occasional hive may swarm. If you also adopt a system of artificial swarming, applied to the strongest stocks as they return from the old seed rape, then there will be no real need to go in for nine-day inspections and the tiresome routine of breaking down queen cells (and possibly missing just one in a corner, so that they swarm after all).

Swarm Catching with Bait Hives

The easiest way of catching a swarm is to provide a place which tempts them so much that they actually occupy it for themselves. Just before a colony swarms, they send out scout bees to look for a new home. If you see inquisitive bees prying into an empty hive, hollow tree or garage space in late May, June or July when no honey is stored there, it means that they are thinking of swarming. When an increasing number of bees are seen doing this in one particular spot, it means that agreement has been reached and that within 48 hours a swarm may be expected to arrive there, not necessarily from one of your own hives. So, preparing a bait hive at the end of May is an excellent way of getting a swarm, without the hassle of climbing ladders and dashing out at short notice after a phone call from the police.

I have found that bait hives are most successful when rather smaller than a normal hive, placed some distance from your own apiary (100 yards or more if possible, but this is not essential) and best downhill if on a slope. The bees also seem to prefer a small entrance, and a cavity well off the ground, at least four to six feet up. I have found that a nucleus box, filled with four or five old brood combs, is very convenient, as the bees readily cluster on the frames and can easily be lifted out and shaken into another hive, so that the same bait hive can perhaps catch more than one swarm per summer. Be warned, if there is a space in the hive not occupied by combs, then the bees will choose to fill it with new comb, possibly built at an angle, rather than use the frames provided. A swarm contains many young bees gorged with honey and programmed to make wax, so perhaps the best compromise is to include two frames fitted with half-inch starter strips of wax foundation, but of course then the bait hive must be levelled so that the frames are plumb vertical.

Hanging Swarms

Usually when a distress call is received via police or council offices, the swarm will be found hanging in a cluster looking like a brown, furry rugby ball, but it pays to ask for the phone number of the householder involved and check a few points before you leave. Make sure that it really is a swarm and not just an unusual number of bees foraging on a cotoneaster bush. Check that they are still there when you set off, and ask for information as to how the cluster is placed and how high off the ground it is, and so on. Try to establish that the swarm is of honey bees and not wasps (black and yellow striped) or bumble bees (larger, rounder and more furry). Neither wasps nor bumble bees form a tight cluster, so this would be conclusive evidence. Talk of many bees just flying around could be a swarm not yet clustered or just many bees gathering honey. To the general public the word 'swarm' simply means a lot of bees. If they are said to be flying in and out of a hole in the ground, they are most unlikely to be honey bees.

Assuming that the swarm is real, perhaps about three and a half pounds in weight (i.e. about 13,500 bees) and hanging from a rose bush about four feet from the ground, use the following technique.

1. Spread a cloth or sacking on the ground over a hardboard square more or less under the cluster.
2. Hold the straw skep or strong cardboard box under and as close to the swarm as possible.
3. Give the branch a sharp shake to dislodge the bees so that they fall in a heap into the box.
4. Quickly invert the box over the sack, placing a pebble or short piece of stick under one edge to provide an entrance.

Some bees will fly back to the branch but the great mass of them, hopefully including the queen, will be inside or at least on the sack, and will be running in. If at this stage, or at any time, the queen is seen, scoop her up into a cage, cork it up and push it under the box. If the queen is inside, a dozen or more bees will soon be seen at the entrance and up to several inches away, heads down and tails up, fanning to broadcast the pheromone with the 'come in' message. At this point, shake the branch again to dislodge the knot of bees re-forming there, and as they fly most of them will pick up the pheromone smell and home in to the entrance also.

Leave them for half an hour, or as long as you have time for, then fold the corners of the sacking or cloth over the skep or box, pick up the hardboard sheet under it and put the whole lot in your car boot. If it is a very hot day and a very large swarm, it might pay to tie the thin rope around the skep to hold the sacking in place, and invert it to provide ventilation through the open weave of the material. If you have only a short drive back (less than half an hour) this is probably not necessary – nineteen times out of twenty it isn't. When you get home or to the apiary where the bees are to be hived, put the swarm box and its contents in a shady spot, propped open a little to allow the bees to fly and to give ventilation, until you are ready to hive them after 5 or 6 p.m. Before you leave, explain to the householder that the next day there may be a fistful of bees clustering on the spot where the swarm was, but that this is not another swarm needing another visit. These will simply be the few bees that were left behind, which will soon disperse and are unlikely to prove a nuisance.

Obviously a brick wall, the trunk of a tree or a telegraph post cannot be shaken, and the bees have to be brushed off, but otherwise the technique is the same. A swarm in a thick hedge can neither be shaken nor brushed – in this case trim off a few twigs so that the box can be placed over the cluster and as close to it as possible, and persuade the bees to run up by gently shaking the hedge, and at the same time puffing smoke at the base of the cluster. Much the same technique may be used for bees on a wall, roof or thick bough of a tree, whenever the swarm box can be placed above and close to the cluster. There are so many variations on this theme that they would fill a book, but such interesting details make excellent bee conversation with colleagues over a drink, and are perhaps best left for such occasions.

Hiving a Swarm

Here there are many interesting possibilities to cover. If the swarm came from one or your own hives, then proceed by preparing a brood box with frames of foundation, including one or two frames of drawn comb, one fairly old and black, and place it on another floor board next to the hive which has swarmed. Also have ready a spare crown board with a side entrance cut into the framing (about 3 × ¼ inches). Wait until early evening, then move the stock which swarmed to one side, sliding it a couple of feet along the stand runners and turning it to face sideways. On the old site put the fresh brood box plus the floor, to simulate the original hive. Now bring up the swarm and shake it into the top of the frames. I usually take out the middle two frames for a few minutes to give the bees room to get down, before replacing them. As soon as possible, but this may have to be half an hour later, encourage the bees still on the top to go down so that the queen excluder can be put on, then the super or supers from the hive which swarmed, plus the bees contained, using a little smoke if necessary. Over the top of the super put the modified crown board with gauze over the feed hole and open entrance to one side, and then lift the old brood box plus bees to go on top, finally replacing the original crown board and roof.

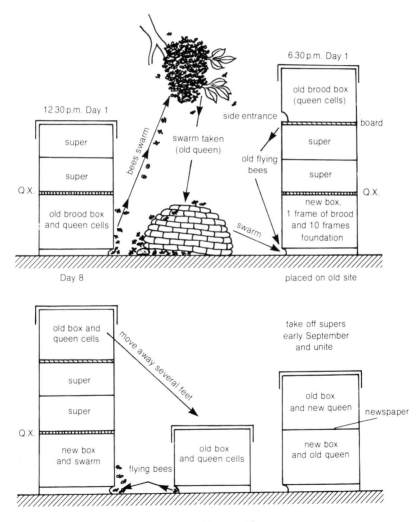

Hiving a swarm from own hive, on old site with original brood box on top.

Summer Management

We now have a situation in which the actual swarm is back on site, with a couple of frames of drawn comb so that the queen can resume laying immediately. Any bees flying from the old stock will return to the old site and be happily received back into the family, to carry on with the good work of storing honey for you, in the same supers. Activity from the old hive above will be little for a few days, but then will build up rapidly as nurse bees develop into foragers and thousands more young bees emerge from frames of sealed brood. Until this happens, convection currents of warm air rising up from the swarm will help to keep the brood warm. After about a week to ten days, or as soon as a considerable volume of flying bees is noticed from the old box above, move this right away on to another floor, give them a feed of sugar syrup, placed over the crown board and under the roof, and leave alone to allow a young queen to mate and take over, building up its strength in good time to get through the winter. The transfer of flying bees lessens the risk of a second swarm or cast emerging, and at the same time boosts the work-force of the swarm, which is where your honey crop from that stock will be.

If you are in doubt as to which hive this swarm came from, capture about a dozen bees from the cluster in a matchbox, drop in a quarter of a teaspoon of fine flour and shake vigorously. Then open the box and watch your hives to see which one the 'white bees' are entering. The flour does no harm and will be combed off as pollen would; the shaking merely disorientates the bees temporarily so they forget they have swarmed. Obviously the hived swarm will still have the old queen, but may well raise a new queen by supersedure. There is also the possibility that a young queen from the swarmed stock at the top of the hive may enter the lower entrance and be accepted, to work alongside her mother before taking over. Should the old queen be retained, it is possibly because the bees are satisfied with her, but there is always the risk that she may fail before next spring. When more experience has been gained, there are several ways of re-queening later in the year.

Stray Swarms

It would be unwise to deny the possibility of acquiring disease from a swarm of unknown origin, but in my opinion the risk is very small, and I have personally not yet met it in over 45 years of swarm taking. Unwelcome characteristics may be present, like bad temper, perhaps a tendency to swarm more than usual, but often good qualities also. A strong swarm early in the year provides the beekeeper with a powerful and willing work-force, and they can always be re-queened with a queen of your own choice if necessary. A normal procedure with a swarm taken from a hedge or tree a mile or so away after a phone call would be to hive it in a spare box fitted with frames of new foundation and feed generously with sugar syrup while combs are being drawn out. The bees are more likely to stay put if you include one frame of drawn comb on which the queen can lay at once.

This is a possible alternative to buying a nucleus of bees in order to get started, and certainly a way of building up stocks when you are in business. I have several times taken two supers of honey (50 to 60 lb) from a strong swarm hived before the end of May, in a good season. Even if you have enough hives of bees already, it is worth while taking a swarm and feeding it well to build up a stock of food combs which will be useful in the autumn, or later that same summer if making up nuclei. The technique is just to go on feeding sugar syrup, and when the bees have frames solid with food at the sides of the hive, to shake or brush them off and replace with frames of wax foundation. The food stored will be a mixture of honey and inverted sugar syrup plus some pollen, very valuable resources to have available. Smaller swarms taken later in

Hiving a swarm collected from elsewhere.

summer will probably be second swarms or casts, headed by young and newly mated queens. Although there will be no prospect of a honey surplus from them in the current year, they can be built up in five-frame half boxes to come through the winter, and in the following year will be strong enough to transfer to full-sized hives and later give a surplus. Alternatively they are there to cover any possible winter losses.

It is important to remember that the wax glands of young bees only go on secreting wax so long as there is a *continuous* income of nectar or syrup. A break in the income for even a few days results in the wax glands shrinking, and after that those bees will produce no more wax, even if feeding is resumed. Thus, unless there is a good natural flow of nectar available, feed continuously and generously – there is no better investment than this.

NUCLEUS MAKING

The easiest method of all is to split a swarmed stock, so that each half has at least one queen cell, and then leave the rest to the bees. Earlier in this chapter I described how the swarmed stock was placed over the hived swarm for a few days before being removed. A variation is to make up two five-frame nuclei on the fifth day, including in each at least one frame with queen cells, and also one frame of food, and move them away to a quiet corner of the garden, restricting the entrance for the first week to a quarter of a square inch. Do not feed for several days as this could set up robbing by other bees, and the only bees in the nuclei will be very young and unable to defend themselves. After about three weeks, or as soon as pollen is seen to be taken in freely, open up to check that the young queen is laying, and mark her.

It is also easy to make up a nucleus from any strong stock in the second half of May and throughout June. Take out two combs of brood plus two outer combs of food and rearrange these in a nucleus box with the food combs at the side. Shake in young bees from another comb, replace the cover and roof and remove to a quiet corner, blocking the entrance with a wisp of grass to hold the bees in for the first day. Do not feed for the first week, as there is a danger of robbing. The technique of shaking in young bees depends on the fact that they cling more tightly to the combs than older bees do, so first shake the frames gently over the main hive, losing most of the old bees, and then shake more vigorously over the nucleus to dislodge the young bees. That frame can then be replaced on the old hive, all the remaining frames pushed into the centre of the hive and frames of foundation added at both ends to make up the correct number.

*Types of nucleus box. The box on the left holds four
frames, the twin nucleus boxes on the right hold five
frames each.*

Before making a nucleus this way it is important to find and isolate
the queen, so that she may remain in the old hive and not be shaken into
the nucleus by mistake. This can be done by putting the frame on which
she is found into a box by itself and covering with a cloth, for a few
minutes until the operation is over. The nucleus has no queen, but does
have eggs, young larvae, food and plenty of young bees. In these
circumstances the bees will construct queen cells around two or three
eggs or very young larvae, and feed royal jelly to the inmates. Swarming
is unlikely, but to be quite certain you can open up after five or six days
and leave just one good queen cell, preferably open, destroying all
others. No further queen cells can be made after this as even the
youngest larvae will by then be too old.

If the strong stock has already put up queen cells, then use combs
carrying them to make up the nucleus and destroy all the others in the
main hive. The loss of bees and combs plus the extra work involved in
drawing out the new sheets of foundation will almost certainly stop
them swarming, at least until they have built up again, so that you will
have achieved two purposes in one operation. There are several more
sophisticated methods of making nucs and raising queens, but these
simple directions are usually perfectly adequate.

Summer Management

When checking hives in mid-July to see that there is enough room for the honey to be stored, you may find that the central combs in the top box are full but two combs at the sides are empty. Rather than add another super so late in the season it may be enough just to move the end combs to the middle, putting the full ones out to each side; bees will work far more readily in the centre combs. This practice cuts down the number of supers on hives to remove at the end of the season, and also gains a few days in which to get ready a further super should it be necessary.

FINDING THE QUEEN

It may be essential to find the queen for a number of reasons. Perhaps a stock has to be re-queened because the old queen is in her third year and slowing down, so that unless she is replaced there will be no honey crop produced. You may be having trouble with an unusually bad-tempered stock, when it will be necessary to remove the old queen before introducing the new one. If the colony is to be artificially swarmed or have a nucleus made from it, it will be necessary to find and isolate the queen first.

It is difficult to find a queen when the hive population is at its peak from about mid-June, but during the first three weeks of April it is usually very easy. To begin with there are less than half as many bees in the hive, and the brood nest (where the queen will normally be found) will probably still be on about five frames, whereas in June there may be twice that number. Perhaps the best reason of all is that the bees are almost always quiet and easily handled on a mild day in April, when there is a larger proportion of young (and gentle) bees. If the search can be timed for the warmest part of the day, early afternoon, then a number of the old bees will be flying and away from home.

The queen will normally be found on a comb with eggs or young larvae, especially if there are also some empty cells ready to be laid in. She is much less likely to be found on a comb of sealed brood, pollen or honey, unless the operator causes a disturbance by too much smoke, or by clumsy handling, when the queen may walk rapidly away from the comb she was laying on and take refuge on outside combs or on the underside of the crown board.

Adopt a stealthy approach when searching for the queen, using a minimum of smoke at the entrance, and ease the crown board up very gently, replacing it immediately with a cover cloth, to exclude light which might drive the queen down to the floor. Always have a look at the bees on the under-surface of the crown board or queen excluder, as about once in 30 or 40 times she will be found there. Go through the hive frame by frame as already described on page 14, but also looking down on the exposed face of the next comb as you lift each frame out. Look for a large, long-legged bee wearing amber stockings! As you lift each frame up to eye level, look first along the lower edge, then round the sides and the top, and finally in the pattern of an imaginary spiral into the centre. Reverse the comb and inspect similarly on the other side. A quiet queen may be seen going about her business of finding empty cells and laying eggs in them, ignoring the beekeeper completely, but a nervous queen may be running around, trying always to get on the other side of the frame, or hiding in any gap between comb and frame.

Marking a Queen

By international agreement, queens hatched in certain years are given certain colours, as shown in the following table.

Year ending 5 or 0   (1990 or 1995) – blue
Year ending 6 or 1   (1991 or 1996) – white
Year ending 7 or 2   (1987 or 1992) – yellow
Year ending 8 or 3   (1988 or 1993) – red
Year ending 9 or 4   (1989 or 1994) – green

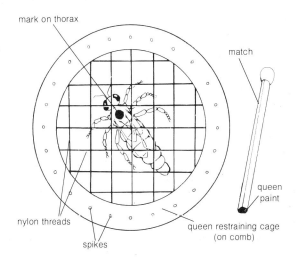

Marking a queen.

However, if your purpose is only to help find her then a spot of white Tipp-Ex or any bright enamel paint may be used. Some beekeepers like to hold the queen but usually the most convenient way is to use a purpose-made queen restraining cage and press this down gently on her while dabbing a spot of colour on the smooth back of her thorax. When doing this for the first time, you might wish to scoop her up gently into a haircurler cage and take her inside. Released by herself on a drawn comb, on a table by the window, she may more readily be restrained without the nuisance of workers getting in the way. Also you can use both hands and work at leisure to make a really good job of it. Non-beekeeping friends or children find this process absolutely fascinating to watch. Having marked her, scoop her up into the cage again and run her in at the entrance to her hive within the hour, or the bees will be worried and running about all over the outside of the hive looking for her.

COMB HONEY

Older customers are always asking, 'Can we have honey in the comb, like we used to get years ago?', and are prepared to pay a high price for it. Most beekeepers do not bother, as in poor summers you get half-drawn or half-filled combs which are unsaleable. In a good year you can do quite well, and the technique is simple. When you have two supers on in June, and the first is nearly filled, replace the second with a comb honey super, with a clearer board between it and the former top one. This crowds the bees down into the comb super – the other one can be completed later on, as a comb super has to be taken off after three weeks to prevent the comb surface being stained and made less saleable. The orthodox comb super is fitted with extra-thin sheets of unwired foundation, but I have always used empty frames with just starter strips of wax and of course for this the hive must be level, at least in the direction at right angles to the frames. Bees build their combs exactly downwards, so if the plane of the frames is not vertical, the combs will appear to be on the slant and not be entirely in the frames.

One useful trick, should you pick up a couple of swarms over a three or four day period, is to run both together into a shallow box of drawn combs, with a queen excluder and comb honey super above, and feed a gallon of thick syrup to get them drawing wax. Alternatively one usually large swarm would serve the same purpose. If there is any honey to be had in the next three weeks you will get 30 lb of comb honey. Then clear the comb super down into a normal brood box placed over the shallow brood box, with no queen excluder, and feed generously with

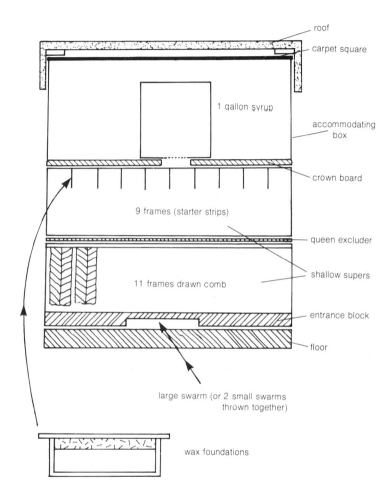

roof
carpet square
1 gallon syrup
accommodating box
crown board
9 frames (starter strips)
queen excluder
shallow supers
11 frames drawn comb
entrance block
floor
large swarm (or 2 small swarms thrown together)
wax foundations

Comb honey production.

syrup to get the stock well established on normal deep frames fitted with the usual wired foundation. By autumn, the bottom shallow box will be empty and clear of brood, and the stock well established in a normal deep box for wintering, ready to give good service next year. Cut comb may be sold by weight in any amounts, either cut out by hand with a sharp knife, or more professionally by using a stainless steel comb cutter to get 8oz combs which fit perfectly into plastic boxes sold for that purpose.

Comb of honey – unwired.

Cutting out 8 oz honeycombs.

HEATHER HONEY

Not everyone will be in reach of heather, but if there is any within two or three hours drive it is worth moving hives there about the second week in August for several reasons:

1   Heather yields a surplus of honey after all else has finished, and so can provide a bonus.
2.   Stocks taken to the heather always build up more rapidly and strongly the following spring, due to the high protein content of heather honey.
3.   Heather honey is the aristocrat of all honies, and justly commands a

Super of heather honey.

Heather press.

higher price, with its golden-amber glow, rich bouquet and superb flavour.

Ideally, for heather you need a young queen in a single brood box, with one or two supers crowded with bees by clearing the normal summer honey supers down into them during the last week of July. As this honey cannot be spun out like other honies, it is best to go for cut comb by using supers of very thin, unwired foundation. Otherwise you must be prepared to cut out and press the combs from your normal supers, which at least gives you a crop of beeswax as well; however, this is tedious and needs special equipment (namely, a heather press). If you tried to get comb honey in June/July and failed because of bad weather, those partially worked combs will be ideal to put on again for the heather. Apart from its unique flavour, heather honey is thixotropic, that is it goes fluid when stirred and then gels again, like some modern paints. The bees did this a long time before the paint manufacturers!

MAKING AN ARTIFICIAL SWARM

This is one of the standard beekeeping techniques, applicable any time between the end of April and mid-July, so long as there are bees working well into the first super and at least seven frames of brood in the brood box. With less than this it should not be attempted. The process is used to make two stocks from one and at the same time prevent swarming – useful if you like to go away for three or four weeks in summer.

Begin by smoking the hive and sliding it sideways on the stand, replacing it with a fresh brood box fitted with wax foundation or drawn combs, of which three have been removed to leave a gap in the middle. Some of the bees returning home with loads of nectar and pollen will automatically enter this hive and begin to cluster. Remove the roof from the old hive and place it upside down between the two hives. Lever the super or supers up from the queen excluder at one corner with the hive tool, and then lift them off and place them diagonally on the upturned roof. Now take off the excluder, and after checking that the queen is not there, prop it up against the hive entrance. Now comes the difficult part, which is of course easier if the queen is marked. Use the technique described earlier to find the queen, and put the comb she is on, plus its bees, into the central gap in the new hive. Look for one other comb of sealed brood plus bees and put this alongside the first. Having made sure that there are no queen cells on either comb, close up the frames, fitting one of the three spare frames in at the end before adding queen excluder, supers and roof.

Artificial swarm, with old brood box on top. (Note side entrance in screen board.)

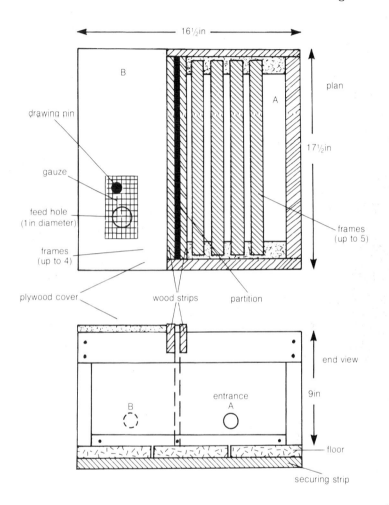

Home-made double nucleus box (construction detail).

Returning now to the old hive, push the frames inward to close the gap before fitting in the last of the two spare frames of foundation at the ends, or better still at the edges of the brood nest. Then replace the super, crown board and roof. Most of the flying bees of all ages will then rejoin the mature queen in the new hive, forming what is virtually a swarm. On the two combs transferred there will be enough nurse bees to look after any larvae present, while the sealed brood will provide about 5,000 young bees to look after the next generation and keep a reasonably balanced population in the meantime, as the queen goes on laying with no break. In the old hive (now queenless), there will be enough non-flying nurse bees to look after the larvae, and queen cells will be built around four to eight eggs or very young larvae. In about two weeks the first young queen will emerge, destroy her sisters in their cells, and given reasonable weather will mate and start laying within the next fortnight. By this time many thousands of young bees will also have emerged from the frames of brood, and there is still plenty of time for the stock to build up to full strength before the autumn.

Frame from mini-nuc (Apidea), with natural comb built down from a 2-inch starter strip.

Wild comb built down from a crown board.

Entrance block opened up for summer.

There should be enough food in the super for the artificial swarm, but the old stock may need feeding as they have to draw out some foundation and to begin with will have no flying bees to collect nectar. If there are any queen cells present, so much the better, but make quite sure there are none on the two combs in the new hive, or there is the possibility of a swarm – though in fact it is more likely that queen cells there would be torn down. If by any chance the queen simply cannot be found, just put any two central combs on which she might be present into the new hive and then proceed as before, and an hour later, look at the fronts of both hives. The one which is queenless will have bees running up and down and around the entrance as if searching for something, and the one with a queen will be behaving normally. If it is the new hive which you judge to be queenless, open up the old hive again and you will find it much less crowded, so that it is easier to find the queen. When found, run her in at the entrance to join the artificial swarm, by holding the frame she is on just touching the entrance and pushing her in the right direction with a feather, before replacing that comb and closing up.

Screen Board/Two-queen Board

An interesting variation is the use of the screen board with gauze panels and side entrance, placed above the artificial swarm and super but below the old brood box. This avoids the need for a separate floor and roof and also enables a two-queen system to be operated later on. Alternatively the top box is an ideal unit for raising queen cells.

SOLAR WAX EXTRACTOR

Although beeswax is dealt with more thoroughly in Chapter 6 (Harvesting the Crop), obviously the solar wax extractor has to be used in

Solar wax extractor.

Summer Management

summer, to melt down the dried cappings stored since the previous autumn as well as any fragments obtained during summer manipulations and hive cleaning. It is usually reckoned that a special factory-made box with expensive (and fragile) double glazing is needed, but I found years ago that a very effective substitute may be made from an old WBC external lift given a substantial insulated floor and fitted with a single sheet of stout plastic (at least $\frac{1}{8}$in thick) which drops into the rebate already present at the wide end of the box. On a clear day in June with several hours of sunshine, even though the air was cool, I have often produced a block of wax weighing 2lb. Even in a poor summer, if left facing south at the right angle, scraps of wax will melt down at no cost or trouble while you get on with other work. The sun also helps to bleach the wax a shade lighter. I have found that one such box will deal cheaply and effectively with all the wax arising from 20 to 30 hives. It is important to appreciate that no worthwhile amount of wax can be obtained from old, black brood combs. This is because the wax present, when it melts, is mostly soaked up by pupa cases and old cocoons left in cells by larvae. Such combs will have to be boiled (*see* page 85). Only combs still showing some yellow colour will melt down in your solar extractor.

6 Harvesting the Crop

REMOVING SUPERS

Apart from the heather, the main honey flows are usually over for the year by the end of July, although in some areas the bees may still be working blackberry and rose-bay willow-herb for another week or so. However, for several reasons I prefer to leave the honey on the hive until the first week in September, apart from hives earmarked for the heather, of course. If you take all the honey supers off at the end of July and then go away for the annual holiday, the bees may starve in a cold, wet and windy August; also much of the honey there at the end of July may have been brought in very recently, and still be too thin for keeping. During August the bees will complete the ripening process and take down any remaining thin honey into the brood nest for current use, or for ripening into winter stores for themselves. Perhaps the best reason of all for a beginner is that the bees are very much better tempered in September, and there are not quite so many of them, so that sometimes the top super may be found completely capped and almost empty of bees. It may then be lifted straight off and a clearer board can be positioned under the rest. My procedure is as follows:

1. Take off the hive roof and place it upside-down on an adjacent hive.
2. Retaining the cover board, lever up honey supers from the queen excluder in the usual way, using anti-crush wedges, and place them temporarily on the roof, diagonally so as not to crush any bees.
3. Put a super of empty combs on the hive, over the queen excluder, then a clearer board over it.
4. Replace the full supers on the hive, over the clearer board, and finally replace the roof. I usually carry a number of carpet squares, cut from discarded material to the size of a hive, and put one of these over the top super. This helps the roof to fit snugly.
5. Check for gaps between the supers, and if any spaces are found, run a strip of masking tape or wide Sellotape around the junctions, to prevent robbing.

All this can be done quickly and without strain, as the heavy supers are kept more or less at the same level, not put down on the ground and

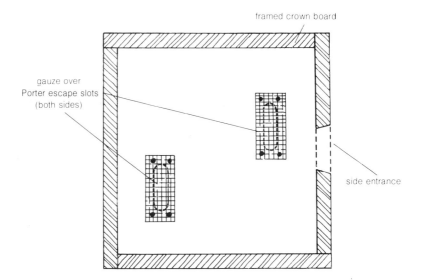

framed crown board

gauze over
Porter escape slots
(both sides)

side entrance

Screen board/two queen board.

Heather super, ready for extraction.

lifted up again. If heavy boxes are lifted from ground level with an arched back, there is a danger of strain and what is known as 'bee-keeper's back'. Usually all the bees are out of the supers after 48 hours, and the honey crop can then be taken off with no disturbance at all to the bees. If you have to carry heavy supers any distance, tuck one side on your hip, with your arm over the top to grip the far side underneath. This way your spine stays erect and the arm only takes half the lifted weight. Sometimes a few bees are still in the supers but these are best ignored – when you have the supers stacked in your shed they will fly to the window and can be released to fly back to their hives, or join up with a hive in the garden and be accepted, as they will be gorged with honey. If Porter bee escapes are left on the hive, the bees tend to deposit propolis on them and gum up the springs, so I usually lever them out before replacing the roof.

EXTRACTING THE HONEY

It is not always possible to extract at once, while the honey is still warm from the hive, and there are considerable advantages in warming up the supers before getting to work. Not least of these is the freedom to leave the stack of full supers a week or two if busy with other work, knowing that 48 hours on the heater pile will bring the temperature back to the 70–80°F (21–27°C) range. If warmer than about 90°F (32°C) the wax cappings will 'drag' on the knife, the combs will lose their strength and be liable to collapse in the extractor. Any cooler than about 65°F (18°C) and the honey is stiff, so that much more tends to be left on the sticky comb surface.

My method of warming up the pile of supers, as shown in the diagram, is very simple and effective. The bridge of thin sheet aluminium or tin plate spreads the heat and prevents the formation of a radiant hot spot directly above the lamp. The empty brood box gives space for hot air to spread out and safeguards the nearest combs from over-heating. The reduced number of frames in the first super also allows hot air to start rising freely and the insulation keeps the warmth where it is needed, in the pile of boxes. Do not expect results in less than 48 hours as the sheer quantity of heat required to warm up seven or eight boxes solid with honey just from 55°F (13°C) to 75°F (24°C) is very considerable.

Electrically heated uncapping knives to use with heated uncapping trays can be obtained and may be used, but I have found that a sensitive palate can detect the presence of a very small proportion of over-heated honey, by its slightly caramelised flavour. For several years now I have

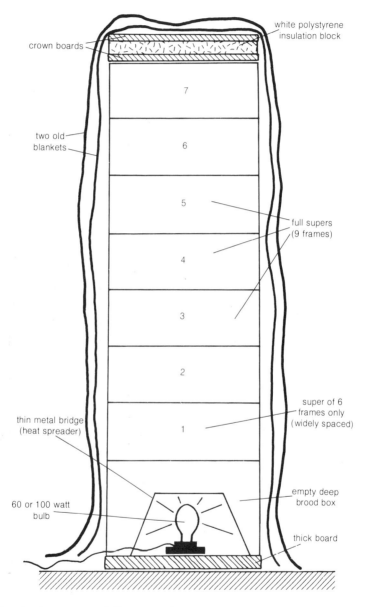

Warming honey supers before extraction.

Uncapping action, with knife.

reverted to my old practice of using a thin, wavy-edged kitchen knife, unheated, in conjunction with a purpose-made uncapping fork. I work with the frame resting on a sieve placed on a two-gallon plastic bucket, slicing under the wax cappings on both sides to leave the cells open, exposing the honey. If the comb surface is uneven I put down the knife for a moment and use the fork on any awkward area. When the sieve appears to be full and clogging I stir up the wet cappings with a round-ended table knife and take a break to let the honey drain through; but if I'm in a hurry I invert the sieve over another plastic bucket, and the whole mass falls away in about half a minute, and the work can go on. Later on the accumulated drained cappings, still containing a lot of honey, can be pressed in a heather press or centrifuged in a purpose-made cappings spinner, which fits inside the extractor. Even at the end of all this, the apparently dry cappings still contain much more honey than one might expect, and it is from this that I make my mead, as described later.

As each frame is uncapped, stand it in the wire cage of the extractor until the full complement of frames is there. If you have a nine-frame radial, then each super provides one load, and of course with a radial the

Uncapping action, with fork.

frames do not have to be reversed. With a four or six-frame tangential, rotate slowly at first and build up to a moderate speed in about half a minute. Then stop, reverse all combs and again rotate slowly at first but build up to a faster speed and continue until drops of honey are no longer heard pattering on the inner side of the extractor. Finally reverse the combs again and run fast for a final minute. These precautions are taken to prevent the full weight of the honey from fracturing the combs when rotating at speed.

As the level of honey in the tank rises, a point is reached where the lugs of frames touch the honey surface; honey must then be run out into containers before continuing. When the last frame has been emptied and cappings centrifuged, leave the extractor overnight propped up at an angle so that the tap is lowest. The last of the honey can be drained out the next morning before hosing clean, drying in sun or greenhouse and wiping with a rag dipped in a little medicinal paraffin before putting away for winter.

BEESWAX

For thousands of years this was the most important of all hive products, until about 150 years ago, it was regularly sold at about eight times the price of honey. In some Third World countries, where bee-hunting is still more widespread than beekeeping, the wild honeycombs taken from trees are crushed with water and the liquid fermented into a strong drink used locally and only the wax was sold. This is because beeswax is valuable enough to stand the cost and trouble of transport along bush paths to town for sale, and durable enough not to deteriorate in extremes of temperature and rough handling. Much of the commercial beeswax sold today is still obtained from tropical Africa.

At home, it is important to collect and save every scrap of burr comb scraped from queen excluders and cover boards, and odd pieces of wild comb built in hives when too large a space has been left between frames. Sometimes when responding to calls about wild stocks of bees in roof spaces or hollow trees, though there may not be much honey there is usually some wax. Make a practice always to have a plastic box labelled 'beeswax', or at least a stout plastic bag, in your bee kit that you always take with you when working.

The bulk of your wax will probably still come from the cappings sliced off honeycombs at extraction time. This can be doubled if you use wide-spaced combs in supers by reducing from eleven to nine frames after the foundation has been drawn out. Then when uncapping cut deeply down to the woodwork and harvest wax as well as honey. This will bring your wax production up from 1 or 1½ lb per 100 lb of honey to nearly twice that figure. The bees will readily extend the cells again next year, and by giving them waxwork to do at a time when under natural conditions they would be doing just that, swarming tendencies are also slightly reduced.

Even when the drained and pressed cappings appear to be dry, there will still be a good deal of honey on them. The only effective way of using this is to soak the cappings in a bucket of cold water, squeezing and stirring until all the honey is dissolved, and then pouring off the sweet liquor (called *must* by wine makers) to use for mead making. After pouring off the must the wet cappings should be pressed by hand and then spread out to dry before being melted down.

Handling Beeswax

While still in the form of combs on frames, or odd scraps kept in bags, wax is liable to attack and destruction by larvae of the wax moth. This pest, which closely resembles its cousin the clothes moth, will lay eggs in odd corners and the tiny larvae will eat their way through the empty cells, leaving a foul mess of excreta in their tunnels. Combs which have been used for breeding by bees are even more vulnerable than clean honeycombs, because their protein content (pupa cases and faecal pellets) supplies the food needs of the wax moth larvae. Wax cast into bars or blocks is not vulnerable to this pest, so it is wise not to postpone the processing of wax scraps and old combs.

Beeswax becomes soft at temperatures of about 90–95°F (32–35°C) and melts at about 145°F (63°C). When overheated it darkens and can become brittle, so is best melted in a water bath. A double pot or porringer is ideal, but a large tin standing in hot water will serve. Prolonged heating should be avoided and it is best to keep the wax only a few degrees above melting point when casting blocks, making candles or wax foundation. Since beeswax has an acid reaction, the natural alkalinity of hard water affects it, producing a surface film of grey mush. Using rain water, or hard tap water with a teaspoon of vinegar per pint, will take care of this problem, which of course only arises when wax is in contact with water. When melted over a water-bath, most of the impurities tend to fall to the bottom, so if molten wax is gently poured or ladled from the top, it will be reasonably clean.

A simple but effective filter may be made from an old pair of tights (fine denier) pulled over a wire coat-hanger bent into an elongated oval shape. This gives a double layer of stretched nylon to pour the molten wax through. As the filter becomes clogged with use, first press down the centre of the area with the rounded end of a thin wooden rod, and

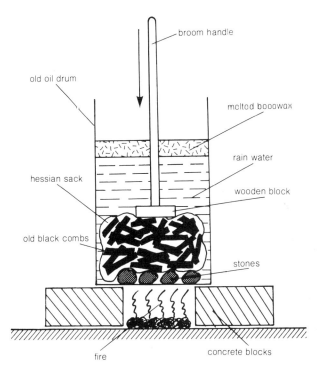

Wax from old, black combs.

then if the wax still does not run through, pull down the far end of the tights to expose a fresh surface.

Uses of Beeswax

Probably most of the wax produced today is used in the manufacture of cosmetics, such as hand and face creams, lipsticks and depilatory wax. Although pure beeswax candles are rarely used nowadays, the best church candles still have a 51 per cent beeswax content. Beekeepers themselves use large amounts for wax foundation, replacing old, gummed up combs with new sheets to be drawn out by the bees. The pharmaceutical industry uses this wax in various ointments, for coating pills and suppositories and so on. Dental wax, used for taking impressions when dentures are made, has a high wax content. Miscellaneous industrial uses are legion, from batik fabric printing and crayons to grafting wax for gardeners, the 'lost wax' process for intricate metal

Harvesting the Crop

castings and of course Madame Tussaud's famous wax models.

I recently ran a bee stall with observation hive in a craft marquee at a fête; on my left was a wood carver stabbing his specialised chisels into a much-used cake of wax (to make them grip the wood, he said), and on my right a leather worker pulling linen thread across a nugget of beeswax to make his thread waterproof, and also stop it from slipping. I supplied both of them with new blocks! Six years ago I sold a two pound block of wax straight from my solar wax extractor to a yachtsman about to sail around the world from Plymouth. Keen fisherman come to me for one ounce blocks for rubbing on their traces and lines. An Australian friend sells beeswax to blondes on the Gold Coast of Queensland; he calls it 'bikini wax'. An antique furniture dealer has a standing order for wax of varying shades from light yellow to almost black; he has seven iron ladles hanging on the wall, each with a mixture of resin and beeswax, providing a complete range of colour to match old furniture. A deep scratch needing to be filled is colour matched, the appropriate ladle warmed over a gas flame and the filling applied while warm and soft; when cold it is extremely hard and durable. A royal chef bakes bread in tins conditioned by rubbing with a piece of beeswax so that the royal bread comes out cleanly. Burglars as well as locksmiths use warmed beeswax to take impressions of keys and locks, and watermarks on bank notes need beeswax in their intricate printing process. Here are just two recipes, though there are many more:

Furniture Polish

Two parts by weight of white spirit to one part of beeswax melted together and well stirred will make an ordinary polish, but a much better one, suitable for valuable antiques, can be made by stirring about a pint (500ml) of warm, natural turpentine into hot beeswax (10oz/300g).

Cosmetic Creams

The basic recipe is 1¾oz (50g) of light coloured beeswax, 6oz (150g) of medicinal paraffin (possibly with a small proportion of almond oil), just under 4oz (110g) of rainwater and ½ teaspoon (3g) of borax, plus a drop or two of perfume if desired. Grate the wax into the warm paraffin in a double saucepan. Dissolve the borax (not boric acid) separately in warm rainwater and stir into the other mixture while both are at a temperature of 160°F (70°C). Continue stirring as the mixture cools, adding the perfume at 140°F (60°C) and pour into warm jars.

POLLEN

To the naked eye the pollen load of a honey bee is just a blob of colourful material about the size of a large match-head, attached to each of her two hind legs. The colours range from slate-blue (bluebells) through golden yellow (dandelion) to brick-red (horse chestnut). Pollen is collected by the bees for its protein content, a necessary foodstuff for tissue building, to enable nurse bees to produce the creamy brood food from their glands to nourish the larvae and to produce royal jelly for the queen, to feed her at all times and particularly when she is laying more than her own weight of eggs every day in summer. Without a liberal supply of pollen in spring, no colony can breed freely and build up the large population necessary to gather a surplus of honey in summer. In the process of gathering pollen, the bees also transfer some of it from flower to flower, leading to the development of fruits and seeds. It is often said that the overall value of this service is several times greater than the total value of all the honey produced.

In a normal year the bees are able to gather plenty of fresh pollen in spring to supplement that stored over winter in the combs, but in occasional bad years (as in 1986), some colonies die out before summer comes, or are so late building up because of pollen shortage that they only just struggle through the summer, with no surplus for the bee-keeper. It is in such a year that the value of feeding pollen or pollen substitute is demonstrated. However, in a normal year there is pollen in abundance in summer, and it is as legitimate to take a surplus of this as a surplus of honey. Stored in a deep freeze it remains viable and can be given back to the bees in February and March should they need it, or sold as a valuable human food.

Collecting Pollen

A professionally-made trap when placed between the hive floor and the brood box, will collect pollen by scraping it off bees' legs as they go through holes in the screen (see page 46). There may very occasionally be some hive debris, like mummified small larvae and the occasional leg, but it is not difficult to pick these out by hand, and the filter grid keeps out any material larger than the actual pollen loads. Some authorities speak of the need to dry pollen very thoroughly, but I find it more convenient to deep freeze it in 2-litre ice-cream boxes. Apart from such pollen, I also get a much larger quantity by scraping pollen-filled cells from honeycombs. The central combs in the first super, just over the queen excluder, usually contain quite large areas of stored pollen, still there after the honey has been centrifuged out. If left in the combs over winter it will go mouldy and be useless anyway, so I scrape down to the mid-rib with an old, sharp-edged kitchen spoon, or grapefruit knife. This pollen, plus the thin side walls of wax cells that come with it, is also stored in a deep freeze until required. Next year the bees will happily draw out comb again from the mid-rib.

There is a distinction between pollen scraped from bees' legs and collected before storage by them, and pollen which has been stored in combs and recovered from them. After storage in cells and having some honey plus glandular secretions added to it, this pollen is technically known as *bee bread*, and medical workers have reported that this is a more effective bio-stimulant than dried pollen. One such report mentioned that bee bread contains an anti-anaemia factor not present in dried pollen. Pollen should never be scraped from brood combs, as their cell walls can contain cocoons, old pupa cases and other debris with a bitter taste. The bees need this pollen for themselves anyway, and it is best not to take either honey or pollen from the nursery department. I usually put back pollen-scraped combs in the centre of the super left directly over the queen excluder for winter. The bees will then tidy up the scraped area and take down any loose pollen remaining.

Pollen stored in the deep freeze can be fed back to the bees in time of need next spring, or more usually added to soya flour before making up the flour/yeast/sugar syrup patties; even a small addition of natural pollen makes the patties much more attractive to the bees. I also sell it to customers as 'natural pollen preserved in honey' by thoroughly mixing one part of pollen with ten parts of warm liquid honey. I make no claim but my customers tell me that if they take half a teaspoon of this mixture with their breakfast cereal each day from the end of January to mid-summer, they do not then suffer from the hay fever which had previously made their lives a misery. Many reports published by Apimondia (the World Bee Association) speak of the value of pollen taken after surgery, in the treatment of chronic alcoholism and other specific conditions, as well as its value as a bio-stimulant or general tonic, even when taken in very small amounts.

COLLECTING PROPOLIS

For medical use, beekeepers scrape propolis off crown boards, queen excluders, frames of honeycombs and the area where frame lugs rest on the rebated surfaces, but never from the floor. Larger quantities may be obtained by placing a sheet of fairly rigid plastic gauze over the tops of the frames for a week or two. When the plastic gauze has been deep frozen for a couple of hours to make the propolis brittle, it will then readily flake off when the sheet is flexed. Since propolis (unlike wax) is heavier than water, a simple flotation process may be used to separate it from most impurities (wax, wood splinters, and bee legs). Break the scraped propolis into small pieces and drop into a bucket of cold water and scoop off the floating bits for the solar wax extractor. Then decant off the water slowly and spread the propolis on a clean tea towel in a dry room for an hour. If a large lump of propolis does not break up readily, leave it overnight in the deep freeze to make it brittle and then crush it. It is well known that bees collect this brown sticky substance from the buds of trees and use it to block up small holes, to stick down frames and any loose parts, even to restrict the hive entrance. It is not so well known that propolis has many other uses, to humans as well as to bees, and is in fact a natural antibiotic.

Uses of Propolis

For Bees

Apart from blocking up any unwanted opening smaller than about that of a queen excluder (0.165 inch, 4.2 mm), propolis is used to varnish all surfaces inside a hive, not only the woodwork of hive body and frames but also the actual waxen cells themselves. The warmth and humidity of the hive environment, with tens of thousands of bees crowded into a small area is the ideal place and conditions for all kinds of bacteria and moulds to flourish. But normally the propolis will prevent this. Poplar and chestnut buds are coated with this brown, sticky substance in the first place as a protection from moulds and infection during the months of winter and early spring. Bees and flowering plants evolved together millions of years before humans appeared on the scene, and each meets the needs of the other, with propolis as with pollen and nectar. The thin varnish of propolis plus bee saliva is obvious on new hive woodwork, but not so obvious on the actual combs; however, tests have established its presence. A mixture of beeswax and propolis is much stronger than beeswax itself, and the bees use the mixture to strengthen their combs at the point of attachment. Also when an intruder too large to move has been killed inside a hive (for example a mouse or large insect) the bees will encase the body in propolis, effectively embalming it so that the flesh does not rot and cause an obnoxious smell, but slowly dries out.

For Humans

In past centuries much use was made of natural balsams such as propolis. Skull trepanning operations, incredibly performed in

Harvesting the Crop

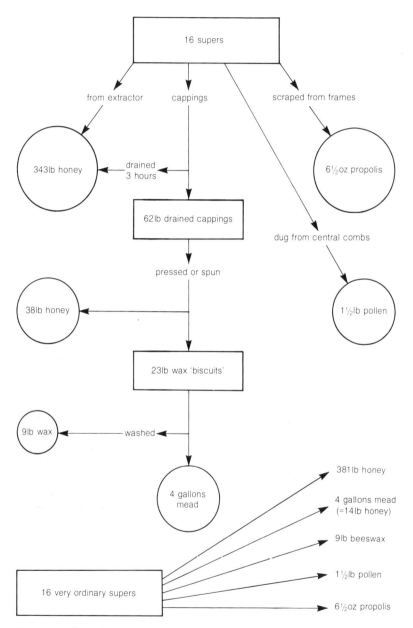

Extraction flow sheet.

prehistoric times, were probably only possible with the aid of propolis as an antiseptic, obtained from wild bee nests. Although looked down on by many doctors as 'folk medicine', there has been a revival of its use, particularly in Eastern European countries and in Scandinavia. Here are a few specific examples of the medical uses of propolis, taken from Apimondia reports of international conferences on the subject held in Bratislava (1972), Madrid (1974) and Bucharest (1976).

1. A solution of propolis in surgical spirit used twice a day as a spray against bed sores.
2. An infusion of crushed propolis in warm water, producing a yellow liquid, as a gargle to relieve a persistent sore throat.
3. Small quantities of propolis (a pinch) added to food and taken daily for infections of the urinary tract, and for prostatitis.
4. A few grams of crushed propolis (about ¼ oz) shaken up in a small bottle with a double tot of brandy or whisky and left for a few days will give a yellow liquid with a strong aromatic smell. Three or four drops of this on a lump of sugar held in the mouth and allowed to dissolve slowly has given great relief in the case of a sore throat.

Violins made by the great Stradivarius were varnished with propolis harvested in the Cremona region of Italy. Recently I sold 4oz (110g) of propolis to a violin maker in Norwich, so evidently the technique persists. Customers sometimes ask what is in this interesting substance – in general terms it is 55 per cent balsams and resins, 30 per cent waxes, 10 per cent ethereal oils and 5 per cent pollen. The balsams and resins are composed of many complex chemicals including flavins, vanillins, and chrysin as well as aromatic unsaturated acids like ferulic acid. Quite apart from the raw materials obtained from tree buds, secretions from various worker bee glands are added to it. For years I have sold it in one ounce packets, with no particles larger than a match head; some take it regularly in very small quantities for intestinal disorders, others keep it for sore throats and suck small pieces. Chemists in several countries (including New Zealand) sell propolis lozenges for the same purpose.

7 Preparing for Winter

WHY FEED BEES?

In Britain we sometimes have summers so poor that many stocks with elderly queens fail to gather enough honey to see themselves through a long winter, even when none is taken by the beekeeper. In nature these bees would perish, and it could be argued that this is nature's way of culling out the poorer strains. This may be partly true, but how drastic! It could be that they tried to replace an old queen in late summer but the new virgin was lost to a bird on her mating flight, or possibly a cold, wet and windy fortnight in July or August prevented the success of the flight. More often, the honey which the bees expected to see them through the winter was in the super taken off by the beekeeper at the end of the season.

It may be argued that honey is always best for bees and that one should always leave a super of honey on the hive for winter, but the truth is that sugar is best for bees in mid-winter, though honey is better in the spring. Sugar is a pure carbohydrate, metabolised to end products of water vapour and carbon dioxide only, with no protein to leave any solid residue in the bowels at a time and temperature when the bees cannot readily fly to evacuate it. Almost no protein is needed in mid-winter, but in the spring it is needed on a large scale for tissue building, and most of the stored food is used to rear young bees. Honey and pollen are then better than sugar, so that feeding sugar syrup in mid-September (two weeks earlier in northern areas), achieves the best of both worlds.

At this time of year the brood nest is much smaller and any incoming food is stored more centrally, so that under the principle of 'last in, first out', it is used first in the November to mid-February period. This leaves the honey and pollen in the outer combs for the critical period from mid-February to the end of April, when protein is essential. In fact, to get an early build-up it is necessary to feed pollen substitute (soya flour/yeast/sugar) in February and March anyway if insufficient natural pollen is available, or if a long, cold spring prevents bees from flying to get it. We need strong stocks earlier nowadays to take advantage of the oil seed rape.

It should be remembered that ordinary white sugar (sucrose), is not stored in this form but converted to a mixture of two simple sugars, (glucose and fructose) by the action of an enzyme called invertase, secreted by worker bees from certain glands. Feeding in September brings the queen back into lay to provide more young bees for the winter, and at the same time the labour involved in 'ripening' sugar syrup shortens the lives of several thousand bees too old to survive right through winter anyway – 'new bees for old', so to speak. In some years there is a considerable income from Michaelmas daisies and ivy from late September almost to the end of October. Ivy honey can crystallise prematurely in the combs, but to an extent this is lessened if sugar syrup is still being ripened and mixed with it.

Sugar Syrup

Some of the older books write of the need to boil sugar and water, even adding cream of tartar. This is quite unnecessary, and it is not even essential to weigh out the sugar. Just tip white, granulated sugar from a 25kg bag into a bucket or drum, note the level of sugar and run in hot water from the household tap while stirring to allow air bubbles to rise.

Making up sugar syrup.

Preparing for Winter

When the sugar/water mixture (the sugar will not yet be dissolved) is up to the same level, the correct mixture for winter feeding has been made (2lb:1 pint, or 5kg:3 litres). For summer use, feeding swarms for example, use slightly more water, which will also speed up the process of dissolving. If you stir the mixture thoroughly about 3 times, at intervals of about 10 minutes, finding some other job to get on with in between, the labour involved in preparing sugar syrup is minimal. The amount of winter food necessary will vary from practically nothing to a great deal, according to the need of each hive, as assessed by hefting and estimating, or by weighing. An average figure would be about 1½ gallons of syrup containing 12 lb of sugar. As a matter of fact this was the amount per hive authorised by the government when sugar was rationed during the two World Wars.

Never attempt to carry syrup in open pails or containers as an emergency stop by your car or van could be disastrous. On a small or medium scale by far the simplest method is to carry the syrup in 1 gallon (4½ litre) plastic screw-top bottles, the sort that supermarkets use nowadays to sell fruit cordials, household bleach, or washing-up liquid. Four of these will sit happily inside an empty super or brood box, to prevent them from falling over when cornering or braking suddenly. Each bottle will hold a gallon containing 8lb of dissolved sugar, and a normal van or estate car can easily take 16 to 20 containers. On a larger

Transporting sugar syrup.

scale I have used 4 gallon plastic jerrycans (used on dairy farms) which fit more easily into a smaller space but are rather heavy for those of us not so young and strong as we were.

Choice of Feeders

Feeding for winter is best done quickly, and the bees themselves are disposed to do this – a really strong stock will readily take down a gallon (4½ litres) of thick syrup, containing 8lb (3½kg) of sugar, in 24 hours. The first point to consider is the practical importance of choosing a feeder to hold at least this volume. All the bee appliance dealers stock one gallon plastic feeding buckets, fitted with a patch of very fine wire mesh in the centre of the snap-on lid. When full of syrup, this is placed over the crown board feed hole with an eke or empty super around it to carry the roof. Sometimes the change in temperature from night to midday is enough to cause the air in a half-empty bucket to expand and make the syrup dribble out, with the risk of robbing if the syrup should trickle out of the entrance. Sometimes up to half a cup of syrup will dribble out when the bucket is first put on, before the vacuum is established; it is wise to invert the full bucket over a spare one first, to catch this. Similar feeders can be made at home from strong lever-lid tins or plastic containers, as described earlier.

In my opinion the best feeders of all are the tray types, such as Miller, Ashforth or Buckfast, which can be kept on the hive for most of the year and be there ready for emergency feeding in early summer, which is necessary sometimes in this climate of ours. Another advantage of this type of feeder is that the odd pieces of comb which bees build in empty places can be removed with a hive tool and scraped off into the feeder for the bees to lick dry, and next time around the dry wax can be removed for melting down without disturbing the bees.

At this time of year bees are very prone to robbing, and any available source of food causes great excitement as they hunt for more; the entrances of other hives are explored and any not well guarded are robbed out. It is now more important than ever not to spill syrup or drop pieces of honeycomb. Avoid the temptation to put out honey-wet equipment for the bees to lick dry. All food should be given privately to the hive concerned, with a well-fitting roof so that only the bees in that hive may have access to it, and entrances should be restricted to make it easier for each hive, especially the weaker stocks, to defend themselves.

Most beekeepers prefer to store empty supers dry, that is to say by having them licked clean by the bees. This is usually done by putting the recently extracted supers back on hives in the home apiary for a week, and can provide a useful addition to the food reserves of hives

Preparing for Winter

Different types of feeder.

concerned. These wet supers should be placed over the crown board, otherwise the bees will sometimes very neatly concentrate the honey (up to $1\frac{1}{2}$ lb per super) in one frame, on the assumption that the super is theirs and will remain in position for their use later on. If it is over the crown board, the bees realise that the food is separated from the brood nest and bring it down into the brood chamber, leaving clean, dry supers of drawn comb for storage in stacks ready for next summer. If you should discover rather late in the autumn (up to mid-October) that a hive has not been fed, or is very light, it is still possible to feed syrup to which a little thymol solution is added, to keep the syrup fresh and prevent fermentation. Any syrup surplus to requirements can also be safely stored in bulk if similarly treated. Thymol may be obtained from chemists in the form of crystals, which dissolve in alcohol but not in water. A stock solution may be made up by dissolving 20g of thymol crystals in 100ml of surgical spirit and adding 1ml of this to every 3 litres of syrup – about a teaspoonful to 3 gallons, but exact dosage is not essential. Such a stock solution will keep indefinitely in a well-stoppered bottle.

MOUSE GUARDS

My own solution to the threat of invasion by mice is to fit a home-made mouse guard-cum-entrance block (*see* page 34) in which a palisade of round nails is driven in at 9mm centres. This totally excludes mice yet offers no impediment whatever to drones, queen or workers laden with pollen, (unlike the official guard with round holes which can scrape some pollen loads off the bees' legs as they scramble through). The advantage of a hinged entrance is that it is never mislaid and can be opened partially or fully from April to July and then pushed in when the honey flow is over and wasps are beginning to become a nuisance; no further visit in October is then necessary to fit an external mouse guard. A final modification is to cut an inch off one of the floor side bars so that the entrance block butts up against it, and cannot be pushed in too far. In wet weather when a block can sometimes swell and jam the entrance, a hive tool can be used to get powerful leverage at one end to open up. This type of entrance block is to a hive what a normal front door is to a house, a far better thing than a loose wooden plank just pushed into the entrance.

LIQUEFYING HONEY

On a small scale it is possible to run warm liquid honey through a conical filter hung on the honey tap of an extractor and bottle directly into one pound jars. It is better to run it into a honey tank (sometimes wrongly called a ripener) fitted with a built-in filter at the top, and bottle off after standing overnight to allow the bubbles to rise. Usually, however, one is too busy at the time and it is best to run honey straight from the extractor into 28lb tins sold for that purpose, or into 30lb white plastic pails, obtained from your friendly local baker, who buys his jam fillings in them. Honey stored in bulk like this normally crystallises after a few weeks and it is then necessary to warm it gently to make it liquefy, in order to bottle it. The problem is that the natural enzymes in honey begin to be destroyed when honey is heated, in a matter of hours above 140°F (60°C) or a matter of days above 122°F (50°C). At temperatures above 150°F (65°C), honey soon starts to darken in colour and acquire a 'cooked' taste due to partial caramelisation. Over-heated honey also suffers a build-up in HMF content to a point above EEC official limit (*see* page 117). So, if honey is to be kept in good condition, high temperatures (even of water well below boiling point) have to be avoided.

Fortunately there is a very easy solution, based on the principle that

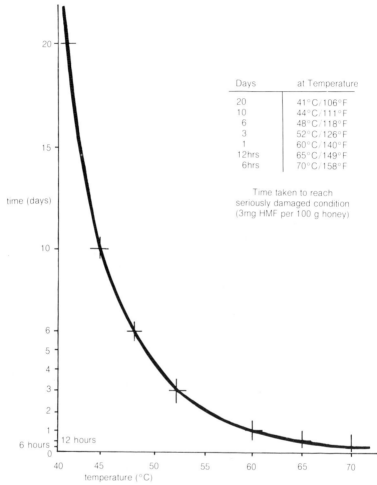

Days	at Temperature
20	41°C/106°F
10	44°C/111°F
6	48°C/118°F
3	52°C/126°F
1	60°C/140°F
12hrs	65°C/149°F
6hrs	70°C/158°F

Time taken to reach
seriously damaged condition
(3mg HMF per 100 g honey)

Honey damage caused by over-heating. (After White, Kushir and Subens, 1964.)

what keeps heat out will also keep heat in. Take an old fridge (easily obtained at no cost - you may even be paid to take it away!) and strip out the freezing box and inside fittings, leaving just a couple of wire racks capable of being slotted in at various heights. Also leave the motor and pump as ballast on the base, to keep the fridge steady. Remove the cooling radiator of metal tubing behind the fridge and thread in a length of flex to take an ordinary bulb holder for the heating element. As the original freezing system holds freon gas under pressure, cut out the tubing when the fridge is outside or where the gas can escape; it is best, though, to have the gas removed professionally. It is neither poisonous nor inflammable but it is still best kept out of the house.

With a 40, 60 or 75 watt bulb lying on a strip of wood on the fridge floor the temperature will build up to about 110°F (44°C) when the surrounding air is in the region of 61°F (16°C). Experience shows that two dozen 1 lb jars will liquefy completely in 12 hours, and two full 28 lb

Honey warming cabinet (2 × 28lb cans).

tins in 48 hours. During winter it may be necessary to use a 100 watt bulb if the outside air is cold, but watch the inside of the fridge carefully if only two or three jars are being warmed up with a 100 watt bulb as the temperature then builds up rapidly and modern fridges have moulded plastic interiors which soften and lose their shape if too hot. One great advantage, so far as warming up large containers is concerned, is that warm honey flows easily through a fine filter into the bottling tank. In fact, because of this, my usual practice is to bottle only my immediate requirements at harvest time, all other honey going straight from the extractor into 28 lb tins for bulk storage until needed.

With a light bulb immediately below a tin of honey, there would be a local hot spot, where honey temperature would build up, so it is best to have a heat spreader made of a 16 × 4 inch (40 × 10 cm) piece of thin sheet aluminium or tin-plate bent to make a bridge over the bulb. As a safeguard against the racks slipping out of the plastic sides when excess heat might soften them, under the weight of 56 lb of honey, I now have short wooden strips to support the racks at each side, just in case, something I did not have to do 25 years ago with the more strongly built, old-fashioned fridges.

Creamed Honey

If a 28 lb tin is warmed for a shorter time, there will be an outer layer of melted honey surrounding an inner layer of warm, soft but still fairly solid honey. This mixture may be creamed by using a metal plunger disc with holes (sold by dealers) pushed up and down vigorously for a few minutes until the contents of the tin are uniformly like very thick cream. This cannot be filtered but may be poured and ladled into jars for sale and is much appreciated by many customers.

MARKETING HONEY

Winter brings more leisure and the time to consider such points as marketing the produce; often there is a peak of honey sales over Christmas, for presents and also to help with coughs and colds associated with cold weather. The great bulk of honey sold in this country is in 1 lb (454g) jars, but also a considerable amount in ½ lb jars. These should have clean caps and be attractively labelled, preferably with the county label. Market research has shown that when shopping, most housewives show a preference for brightly coloured labels, for example of a thatched cottage with attractive flowers. Strange though it may appear to beekeepers, it seems that such customers dislike pictures of

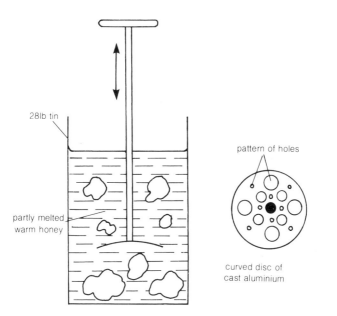

Creaming honey.

honey bees but react favourably to pictures of large, furry bumble bees! Most customers prefer clear, run honey but a minority also demand granulated honey; a soft, creamed honey sells very well. Comb honey with a clean surface, presented in 8 oz white plastic containers with transparent lids, will sell readily at a price approximately twice that of ordinary honey. Over a few years most beekeepers build up a good volume of sales direct to the public, to regular customers who return the empty jars, an important consideration when they cost over 10p each, even when bought by the gross. Regular sales may also be achieved via WI markets or from a bee stall with observation hive at a fête or craft market. Many local organisations such as Probus clubs, church groups, and women's associations are always asking for speakers to give talks on bees, and usually ask if they may buy honey direct from a beekeeper. Local shops are usually very willing to take two or three dozen jars at a time at an agreed discount.

8 Pests and Diseases

With all farm stock, the maintenance of healthy vigour is better than treatment, and this is certainly true with bees. The most important single factor is sensible management: making sure that hives are sound with waterproof roofs, fitted with mouse-proof entrances in winter and restricted entrances in wasp time (August to October), and that the bees have plenty of food on the combs (at least 10 lb) at all times. Maintain a good standard with your brood combs, culling out two of the poorer ones from each box every year after the second. Routinely sterilise brood combs not in use with 80 per cent acetic acid in April. Be content with only two or three hives in a poor area. In a really good area where up to 20 hives may be kept on one site, see that they are not arranged in straight lines, but disposed in irregular groups of 3 or 4 to cut down drifting, whereby foraging bees accidentally enter (and when carrying food are accepted into), hives other than their own, spreading disease should there be any.

A gas blowtorch is essential, to sterilise by scorching any floors, roofs and boxes when not in use. Pay particular attention to corners and crevices where wax moth larvae may be hiding. In all your manipulations take great care not to crush a bee: remember that the only way the house bees can clear up a squashed bee is by licking up the body fluids before dragging away the pieces, and that one bee with nosema spores could infect 50 bees in this way. The Ministry of Agriculture at York operate a diagnostic service for adult bee diseases, so if in doubt, send a sample of about 40 to 50 recently dead bees in a matchbox, with details of the hive they were taken from – do not wrap them in plastic as the specimens will quickly rot. Many counties and local associations also make arrangements for this work to be carried out at a local Agricultural College or by experienced volunteers. In cases of suspected brood disease, contact the local Bees Officer; these usually work from 1 April to the end of August only, the active period for brood rearing. Hive debris found on floors in spring should also be tested routinely, either at York or locally, as a check on varroa. If spray damage is suspected, pack several hundred dead bees in a ventilated box and despatch to York immediately, with a brief note of the circumstantces and details of crops grown locally which may have been sprayed.

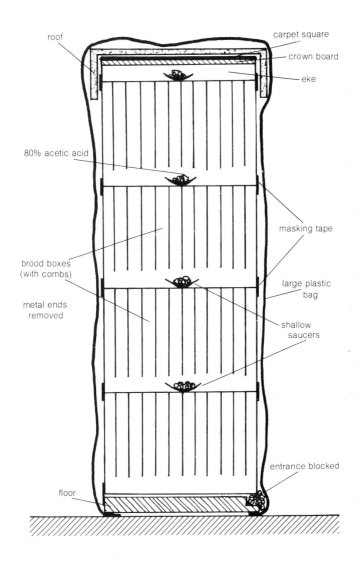

Sterilising brood combs.

Scorching floors and brood boxes.

ADULT BEE DISEASES

Nosema

This may be recognised by the accompanying dysentery and visible staining of combs and hive fronts, but failure of a colony to expand in spring and numbers of crawling bees in front of a hive are also symptoms. The organism itself (*Nosema apis*) multiplies in the lining of a bee's gut, seriously weakening the affected insect and shortening its life, usually without killing it directly. Insects cannot regenerate body cells once they have been destroyed. The organism multiplies by producing millions of spores (like seeds) which germinate when they reach the intestine of another bee via food from infected combs, or from another bee. This disease is as endemic to bees as the common cold is to human beings, and only becomes a serious problem when stocks are sited in poor areas, over-manipulated or subjected to stress of some kind or another. Bees that have been weakened by some other disease, like paralysis or acarine, or have insufficient food or a queen of poor quality, are more likely to suffer. Strong, well-fed stocks with good queens will throw off nosema, yet may still have traces of the infection in just a few bees.

Clinically-minded beekeepers may confirm the presence of nosema by examining the large intestine of a bee, which will be much bigger than normal and white instead of very pale brown. Those with a microscope may examine an intestinal smear under ×400 magnification, preferably with a droplet of a dark stain (nigrosin or methylene blue) which helps the nosema spores to be seen, like grains of rice. Routine sterilisation of combs with 80 per cent acetic acid and thorough scorching of floors and empty brood boxes will do much to keep down infection. When nosema is clearly identified there is a specific remedy – Fumidil B. This is an antibiotic with no other use, produced just for this purpose and obtainable through local associations or bee appliance dealers, with full instructions for use. Obviously the antibiotic is for use on hives of living bees, and the acetic acid and blow torch reserved for equipment with no bees present!

Acarine

Also known as 'Isle of Wight Disease', this is caused by a small mite which breeds and lives in the large trachea or breathing tubes of adult bees. These parasites suck the haemolymph (bee blood) via small holes made by their prosbosces, killing or seriously weakening the affected insects. As well as direct damage so caused, the physical presence of large numbers of these parasites blocks the breathing tubes and reduces air circulation; also secondary infection by some virus may enter via the wounds in tracheal tubes.

The symptoms may be obvious, with numbers of bees on the ground crawling away from the hive, or climbing up grass stems and unable to develop enough power for take-off. Some may be 'K-wingers', with wings distorted and held open at an angle like a capital K. I have noticed that in early spring an infected colony is restless and by listening close to the brood chamber on a quiet early morning or evening brittle, crackling sounds may be noticed. This parasite (*Acarapis woodi*) cannot live for more than a few hours away from bees, and the trouble cannot arise from infected combs or boxes. It is passed from hive to hive by bees, either transferred on combs by the beekeeper or drifting into another hive after flying.

In my experience the most effective treatment is by Folbex strips in spring when you are most likely to become aware of the problem and no

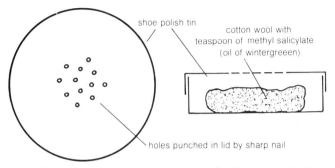

shoe polish tin

cotton wool with
teaspoon of methyl salicylate
(oil of wintergreeen)

holes punched in lid by sharp nail

A method used to prevent acarine. In September the flat tin is pushed into the entrance to lie centrally on the floor beneath the brood box during the winter, and removed the following April.

honey supers are on the hive. These strips, obtained from your association or any dealer, are made of thin green cardboard which has been soaked in chlorobenzilate (the active principle) plus a little saltpetre to make the material smoulder slowly until burnt out. Using a drawing pin, fasten a strip to the inside of the hive roof so that it hangs down over the feed hole of a crown board placed over the brood chamber, with an empty shallow box between crown board and roof. In the evening, when all the bees are at home, block the entrance with foam rubber, ignite the end of the Folbex strip and when it is smouldering with a trail of smoke, replace the roof. The heavy smoke will then drift down into the brood area. After an hour, very quietly remove the foam rubber from the entrance. It is recommended that a second treatment be given a week or ten days later. I remember one colony which I treated several years ago, recovering so completely and later in summer doing so well that I almost forgot it had ever been treated, until I noticed the burnt-out end of a Folbex strip on the inside of its roof. Recently it has been alleged that Folbex is carcinogenic, and the strips are difficult to obtain, but oil of wintergreen is still readily available.

Dysentery, Amoeba and Paralysis

Dysentery is really a symptom rather than a disease, and may sometimes indicate nosema. Sometimes it is noticed after a long, cold spring when bees have been unable to fly and evacuate normally. Amoeba affects the Malpighian tubules (equivalent to kidneys in animals) and can often cause dysentery. Paralysis is due to a viral infection and causes the bees to tremble, often giving them a greasy, bald appearance. At present there is no known treatment for any of these, other than to re-queen and hope that the new stock will have a natural resistance to the trouble.

BROOD DISEASES

Chalk Brood

Caused by the fungus *Ascophaera apis*, chalk brood is perhaps the most common condition, and in early summer a small number of hard, white, mummified larvae may be seen, especially near the edge of the brood area. The condition may be distinguished from chalky, mouldy pollen (which it closely resembles) by probing, as old pollen crumbles but chalk brood does not. Most colonies show traces of this if one looks closely enough, but serious trouble only arises (if at all) when the colony is weakened by some other condition. There is no specific remedy – good management plus feeding or re-queening where necessary are recommended.

American Foul Brood (AFB)

This is the most serious disease affecting larvae, which collapse and die *after* cells have been sealed; their colour changes to a creamy brown and finally to very dark brown, almost black scales lying on the lower side of the cell from just behind the cell mouth right back to the base. The cell cappings become moist and collapse inwards as the larvae shrink. In their attempts to remove the larvae, the bees may nibble holes in the cappings and sometimes chew them away completely. The organism is *Bacillus larvae*, highly dangerous because it forms millions of very resistant, oval spores which can remain viable for many years. Infection is via these spores which germinate in the gut of very young larvae.

The disease is spread by a strong stock of bees robbing out a weaker, diseased stock; by nuclei made up with combs from infected stock; by the use of old combs bought secondhand; by infected honey (imported) left in tins or drums and cleared out by bees; and by young bees from an infected hive drifting into other hives after orientation flights. Where the disease has been suppressed by antibiotics (usually terramycin) hives may still produce a surplus of honey, but remain reservoirs of infection if not continually treated. The only effective remedy is destruction by fire, with compensation by insurance from BDI, included automatically for the first two colonies in your subscription to your local branch affiliated to the BBKA. So far as the beekeeper is concerned, AFB is a notifiable disease and must be reported to the local Bees Officer, usually referred to as the Foul Brood Officer, who will inspect immediately and take all necessary action. The usual home diagnosis is by probing affected cells with a matchstick. If a ropy string of brown matter (with a foul smell) appears when the match is slowly removed, AFB is indicated.

European Foul Brood (EFB)

This also affects larvae, but death occurs *before* cells have been sealed. The actual cause of death is starvation, as the organism *Mellissa coccus pluton* competes for food in the larval gut. The larvae lie uneasily in their cells at unusual angles and die when four or five days old. After death they turn brown and decompose, giving rise to a sour smell. No highly resistant spores are formed and dried up cocci survive no longer than about three years. Infection is from other hives, usually in areas where the disease is endemic, from faecal remains on combs. Again, this disease is notifiable, and the local Bees Officer may administer terramycin with your agreement and at his discretion.

Sac Brood

This is a viral disease which attacks the brood in the pupal stage, forming tough little sacs filled with fluid. A good honey flow usually clears it up, and conversely it seems to flare up with stressed colonies. There is no cure, other than to re-queen.

Chilled Brood

This may be noticed as grey and undersized, dead larvae, especially at the edge of the brood nest after a sudden spell of cold weather has caused it to contract. The bees will soon clear this up when the weather improves.

Addled Brood

This is usually attributed to a genetic factor, or a virus of unknown origin. Sometimes a very small patch of dead larvae can be attributed to minor spray damage, for example, an enthusiastic amateur spraying his two apple trees half a mile away. The bees normally clear up these symptoms as summer develops, and often they pass unnoticed.

Spray Damage

Although not a disease, this can be deadly when chemicals like chlorinated hydrocarbons and other pesticides are used on a large scale by farmers on crops such as oil seed rape. Farmers are aware of the problem, but sometimes employ contractors who are less interested and use more deadly formulations than expected. Removal of hives near the end of the flowering period, or partial obstruction of the bees by forking hay over the hives will check full-scale foraging for 24 hours but there is no complete solution to this problem, other than 100 per cent liaison and co-operation between farmers and beekeepers.

Braula Coeca

Infestation by this insect rarely causes serious trouble, but can be a nuisance to a queen. Also, in the larval stage, this pest tunnels under wax cappings and can make comb honey unsaleable by surface damage and disfigurement. *Braula coeca* is a tiny, flightless reddish-brown insect with six legs. It is not a parasite and causes no damage to bees, but helps itself to food from a bee's mouth parts. The insects are agile and can hop like fleas, but usually cling tightly to a bee's back. Tobacco smoke will dislodge them, so the recommended treatment is to roll up some cigarette ends in corrugated cardboard and use the roll in the smoker, with a sheet of stiff paper on the hive floor to catch them when, stupified by the tobacco smoke, they fall off the bees. Remove the paper after half an hour and destroy, before the creatures recover and get back on to the bees.

Varroa

The presence of this mite in the UK was first identified by the author and three colleagues in Torquay on 2 April 1992, during a training exercise organized by the local branch of the Devon Beekeepers' Association for that purpose, but with no expectation of finding anything. For the record, it was Margaret Saffery who first noted the unusual shape and said 'What is this?' Within days the BBC were at Cockington Apiary to record for television the first varroa mites seen in Britain. The infestation discovered was very slight, but within a few weeks much more heavily infested stocks were found at a number of sites along the south coast. With the benefit of hindsight it is clear that varroa had been in the country for three or four years at least. Perhaps the most dangerous aspect of this problem is that there are no obvious signs of trouble until about two or even three years after the first mites have arrived (via drifting workers or drones), by which time the colony may be near total collapse. For this reason it is necessary to take positive steps to check whether a colony is infested or not.

The scale of infestation can be assessed by using a floor insert of greased card and counting the natural drop of mites after a couple of days. A purpose-built varroa floor helps but is not essential; if you have one, check it regularly to make sure it is not a refuge for wax moths.

Varroa is an external parasitic mite which breeds on bee larvae,

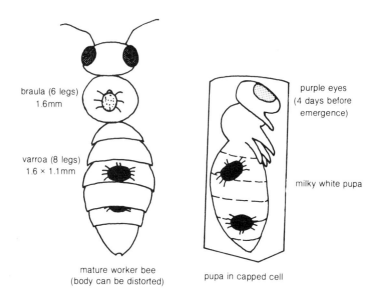

braula (6 legs)
1.6mm

varroa (8 legs)
1.6 × 1.1mm

mature worker bee
(body can be distorted)

purple eyes
(4 days before
emergence)

milky white pupa

pupa in capped cell

Braula and varroa.

sucking their lifeblood (haemolymph); this is where most of the damage is done. When not in brood cells the mites live on bees in the hive, but in summer about 85 per cent of mites at any one time are probably on larvae or pupae in cells. Normally the mother mites enter brood cells some hours before they are capped.

A further complication is that wounds inflicted on larvae and bees themselves by feeding mites can permit the entry of viral or bacterial infection. Such secondary infection, such as bee paralysis, can be more lethal than the mite itself.

At the time of writing the only treatment licensed in the UK is Bayvarol, manufactured in strips containing flumethrin, a synthetic pyrethroid. To avoid any possibility of chemical residues in honey, these are used only when no honey supers are present on hives. It is essential that all stocks in an area are treated at the same time, so co-operation with local beekeepers on timing has to be arranged, usually in August. Strips have to be removed after six to eight weeks. Many find it convenient to use long-nosed pliers to get a firm grip when bees have stuck the strips in with wax or propolis.

After hiving a swarm, the time interval can be reduced to a few days, as for the first week any mites present can only be on the bees them-

Grade of infestation	No. of dead mites per day	Infestation on drone comb	Infestation on worker comb	Infestation on adult bees	Observed colony behaviour
Light	0-4	Occasional	Not noticed	Not noticed	Normal
Medium	5-10	Clearly visible	Seldom noticed	Not noticed	Normal
Heavy	11-15	Almost everywhere	Clearly visible	Sometimes noticed	Fairly normal
Critical	16+	Complete	Widespread, looks like foul brood	Crippled, stunted bees seen	Irritated. Bees no longer clustered on brood

A reasonable assessment of the degree of infestation may be obtained by placing a clean oiled or vaselined sheet of paper on the hive floor and withdrawing it 24 hours later. Beekeepers in Germany have linked the normal daily mortality of varroa mites as counted, with the degree of infestation as shown in the table above. (Courtesy: Professor Wolfgang Ritter)

selves and are immediately vulnerable. In such cases two strips instead of the usual four would be adequate. Normally one treatment per year is sufficient, but where re-infestation from wild colonies (or non-treated stocks) is possible, it may be necessary to re-treat again in February. New beekeepers are advised to consult their colleagues in local associations on points such as this.

Laying Workers

Although not a disease, this condition is as awkward to handle as any pest. It only arises when a stock of bees has been queenless for several weeks, usually in summer, when a number of worker bees develop their normally immature ovaries until they can each lay a few eggs a day. These are infertile and result in undersized, useless drones. The condition may be recognised by noting more than one egg in a cell, typically deposited on the sides rather than the base. Probably the first symptom to be noticed will be patches of high-domed drone brood scattered around the brood nest, in the absence of a queen and normal brood. At a later stage numerous undersized drones will be seen flying or walking on the combs. The only other occasion on which several eggs may be seen in each of a number of cells is when a newly mated queen has insufficient room to lay, and in her youthful enthusiasm lays more than once in several cells. However, all but one egg per cell will be eaten by the bees, and normal worker brood produced. If the population is small the colony is probably not worth saving, and the best solution is to move the hive a few feet from its previous position and block the entrance; then shake out all the bees into the centre of the apiary, when most will join up with other hives and do useful work. Any attempt to introduce a new queen or queen cell will almost certainly fail. One certain cure is to shake in a swarm, which will then take over by sheer pressure of numbers and rapidly tear down the undersized drone cells in an orgy of spring cleaning. The laying workers then either revert to normal or are eliminated and the bulk of the stock then happily reinforces the swarm.

Wax Moths

These resemble their cousins the clothes moths, being about half an inch in length, usually of a drab, greyish-brown colour with a silver sheen. When disturbed they run rapidly about on floorboard or comb with a deceptive zig-zag avoiding action that makes it difficult to catch or jab them. They can squeeze through very narrow cracks. It is the larvae that do the damage, as they tunnel through wax combs, eating some wax but spoiling much more as they protect their tunnels with a web of silk reinforced by their excrement, making it difficult for bees to get at them. In extreme cases a box of combs can be reduced to a foul tangle of criss-crossed tunnels with hundreds of pupae lying flat on the frames or walls of the box. The pest can actually eat into the woodwork, making shallow, oval depressions in which the larvae pupate. Fortunately this pest is extremely vulnerable to cold, and supers of extracted honey-combs, licked dry by the bees as already described, may be stacked outside for the winter, with queen excluders top and bottom and of course a sound roof. Usually a friendly garden spider finds its way into the pile and exercises biological control. On a small scale, one night in a deep freeze will kill all stages of this pest. Some authorities recommend crystals of PDB sprinkled in each super before wrapping in newspaper. Do *not* just wrap supers in newspaper alone and store in a spare room in a centrally heated house, or you will receive a nasty shock six or seven months later when you unwrap them!

9 Legal Aspects

WHAT CAUSES PROBLEMS?

An analysis of a large number of case histories over several years was most revealing. More than half of all complaints were 'frivolous and vexatious', usually when neighbours had previously quarrelled about something else, like a loud radio in the garden, children's footballs going over the wall, motorbikes revving up on a Sunday morning, smoke from garden fires and so on. Of the genuine grievances, about three out of five were concerned with stings, either actual or sometimes just the fear of them. About one in five were to do with swarming, sometimes true but often just a large number of bees foraging on a cotoneaster bush, for example. The remainder were mostly about 'brown spots' on the washing or on newly polished cars, and humorous remarks about highflying aircraft and 'blame British Airways' did not help. When bees fly on their so-called cleansing flights, especially in spring, they may deposit faecal matter closer to the hive than in summer, when the weather is warmer and they fly further afield. Young bees need to eat pollen in order to produce brood food, and the outer parts of pollen grains cannot be digested, so that this adds to the problem of waste disposal.

Very rarely are there complaints about farm stock being stung or affected in any way. Animals soon learn not to go too close to beehives, unless of course they are tethered or confined very close to an apiary. In one rare case, a farmer who, in 1906, established an apiary of 16 hives on the border of another farmer's land close to his stables, with the result that two horses were so severely stung that they died, was found guilty of 'negligent and inconsiderate disregard of the duty of one neighbour to another' and substantial damages were awarded. Tractors have now replaced horses, but cattle, especially if confined in a yard close to an apiary, could possibly be the subject of a similar legal action.

Swarms of Bees

Swarms have been the subject of myth and folk lore, of 'tanging' and uninformed comment that 'a man may follow his swarm anywhere' under a law said to have been enacted by Alfred the Great. In truth the

Legal Aspects

legal position has varied little since Roman times. Bees are the private property of the owner of the hive only so long as they remain tenants of the hive, when the unauthorised removal of them constitutes theft. But if the bees issue from the hive as a swarm they are deemed to have reverted to their wild state and become *ferae naturae* when they have escaped from the sight of the owner, or have proved too difficult to recapture. The right of property is then vested in whomsoever captures them. In this respect bees, pigeons and deer were regarded as 'wild creatures temporarily owned' as opposed to such domesticated animals as cows, horses and sheep. The Roman law of Justinian has, in practice, been modified by the acceptance among beekeepers themselves that ownership of a swarm should be respected if it is clear whence it originated. There is no right of entry to the property of another person in order to take a swarm but usually such action is welcomed. At the worst, the trespasser faces an action for damages, and a reasonable person quietly taking a swarm, being careful not to trample on a growing crop or damage a fence has little to fear. The doctrine of 'hot pursuit' could well be quoted in mitigation.

Bees and Neighbours

In the absence of specific legislation, situations in Britain are governed by common law and precedent. The basic concept is of a 'reasonable person' and what such a person might be expected to do. Anyone who keeps bees of such a nature, or in such numbers, or in such a place, or in such a manner as to deny a neighbour his normal and reasonable enjoyment of house, garden or property, may be committing a nuisance. Although there is no relevant statute regarding bees, there is the Public Health Act, 1936, Section 92 of which refers, at paragraph 1b, to '... any animal, kept in such a place, or in such a manner as to be prejudicial to health or cause a nuisance ...' In the explanatory paragraph it is stated that in this connection poultry and bees are regarded as animals.

A study of actual cases over the last 100 years reveals that in practice, *scientia* or alleged knowledge of propensity to sting, has rarely succeeded as a ground for action. Rather surprisingly alleged negligence has also brought very few convictions, although it may be a contributory factor to 'causing a nuisance', on which almost all successful prosecutions have been based. An isolated example of stinging has not usually been upheld as causing a nuisance, and it has to be established that the offence has been of a sustained kind. Where the defendant has demonstrated his willingness to take some remedial action promptly, such as moving hives away, the case has usually been dismissed.

BEE DISEASE LEGISLATION

The Importation of Bees Order, 1980

This prohibits the import of bees into Great Britain except under an import licence, which has to be issued by the Ministry of Agriculture in respect of each consignment. A health certificate completed by the relevant officer in the country of origin has to accompany the bees. If a queen bee is to be imported her attendant workers must be sent to the Ministry at York for examination. A bulk importer of queens must replace attendant workers before distributing queens to customers; the importer must also keep a record of the ultimate destination of each queen.

The Bees Act, 1980

This gives power to the MAFF to make orders to prevent the introduction and spread within Britain of pests and diseases affecting bees. Contravention of such an order may lead to a fine of up to £1,000.

The Bee Diseases Control Order, 1982

Under this act, any beekeeper who has reasonable grounds to suppose that his bees may be infected with AFB, EFB or varroasis is obliged to report to the proper authority, and must not move his bees or any equipment, including honey and bee products, until so authorised. In his official capacity the Bees Officer, usually referred to as the Foul Brood Officer, may take samples, mark any hive or equipment and issue a 'standstill order' when he has reasonable grounds for suspecting disease, or is refused entry to a place where he believes bees, equipment or produce are present.

AFB Confirmed

If laboratory tests prove positive, a notice will be served requiring destruction by fire of all bees, combs, honey and quilts. The beekeeper is also given the option either to destroy by fire, or treat all other equipment which appears to the authorised person to be infected or to have been exposed to infection. Where a beekeeper agrees by signing the order, an authorised person may serve a destruction order without waiting for a laboratory test.

115

Legal Aspects

EFB Confirmed

In this case, a very similar notice will be served, except that there is the option of treatment as an alternative to destruction. Where bees are treated, they must remain in the custody of the beekeeper for eight weeks and no combs or products removed during that period except under licence.

Varroasis

If varroasis is confirmed, an order may be served requiring the destruction, treatment or isolation of the hive, bees and appliances which have been affected. In addition, an area may be declared as infected, within which an order may be placed requiring the destruction, treatment or isolation as stated above. Any movement into, within or out of the area will be prohibited. In all cases the notice will give details of methods of destruction or treatment and timing, and whether these are to be carried out by or in the presence of the authorised person.

Other UK Acts and Regulations

Under the Public Health Act, 1936, trade in honey in the UK is governed by the following:

Food and Drugs Act, 1955
Labelling of Food Regulations, 1970 (as amended)
Honey Regulations, 1976
Materials and Articles in Contact with Food, Regulations, 1978
Weights and Measures Acts, 1963 to 1979
Weights and Measures (Marking of Goods and Abbreviations of Units) Regulations, 1975 (as amended)
Weights and Measures Act, 1963, (Honey) Order 1976
Trade Descriptions Acts, 1968 and 1972
Trade Descriptions (Indication of Origin – Exemption 1) Directions, 1972
Consumer Safety Act, 1978
Glazed Ceramic Ware (Safety) Regulations, 1975

DEFINITIONS OF HONEY

Honey is the fluid, viscous or crystallised food which is produced by honey bees from the nectar of blossoms or from the secretions of, or found on, living parts of plants other than blossoms, which honey bees collect, transform, combine with substances of their own, and store and leave to mature in honeycombs.

Comb Honey is honey stored by honey bees in the cells of freshly built broodless combs, and intended to be sold in sealed whole combs or in parts of such combs.

Chunk honey is honey which contains at least one piece of comb honey.

Blossom honey is honey produced wholly or mainly from the nectar of blossoms.

Honeydew honey is honey, the colour of which is light brown, greenish brown, black or any intermediate colour, produced wholly or mainly from secretions of, or found on, living parts of plants other than blossoms.

Drained honey is honey obtained by draining uncapped broodless combs.

Extracted honey is honey obtained by centrifuging uncapped broodless honeycombs.

Pressed honey is honey obtained by pressing broodless honeycombs with or without the application of moderate heat.

EUROPEAN HONEY LEGISLATION

Composition of Honey

1.   There should be no addition of substances other than honey.
2.   The honey should be, as far as is practicable, free from mould, insects, insect debris, brood and any other organic or inorganic substance foreign to the composition of honey. Honey with these defects should not be used as an ingredient of any other food.
3.   The acidity should not be artificially changed and in any case must not exceed the legal maximum level of acidity.
4.   Any honeydew honey or blend of honeydew honey with blossom honey should have an apparent reducing sugar content (invert sugar) of not less than 60 per cent and an apparent sucrose content of not more than 10 per cent.
5.   In general, honey shall not contain more than 21 per cent water, but heather and clover honeys may contain up to 23 per cent water.
6.   HMF content must not exceed 40mg per kg (40 p.p.m.) At the moment the limit for UK sale is permitted to be twice this.
7.   Enzyme content, measured in terms of diastase activity, must be not less than 8 (4 at present for UK sales).

Legal Aspects

Bakers' or Industrial Honey

Honeys of the following descriptions should be labelled or documented only as bakers' honey or industrial honey.

1. Heather honey or clover honey with a moisture content of more than 23 per cent.
2. Other honey with a moisture content of more than 21 per cent.
3. Honey with any foreign taste or odour.
4. Honey which has begun to ferment or effervesce.
5. Honey which has been heated to such an extent that its natural enzymes have been destroyed or made inactive.
6. Honey with a diastase activity of less than 4, or an HMF content of more than 80 mg per kg.

MARKETING HONEY

Honey in pre-packed form may only be sold in the UK in certain specific quantities: 1 oz, 2 oz, 4 oz, 8 oz, 12 oz, 1 lb, 1½ lb or multiples of 1lb. Chunk honey and comb honey can be packed in any quantity. Honey sold wholesale in containers of 10kg and above need not be labelled, but must be accompanied by a consignment note giving all the information which would otherwise have been included on a label. Honey for retail sale should be marked with the following information:

1. An indication of quantity by net weight in both imperial and metric units, in figures at least 6mm high, so far as 1lb and ½lb packs are concerned.
2. The name or trade name and address of the producer, packer or seller.
3. A description of the contents, whether honey, chunk honey, comb honey, heather honey, Devon honey etc, but whatever the description, it must be true. There are two basic types of illegal mis-description: the direct type such as describing Mexican honey as Devon honey, and the indirect or misleading. An example of this would be an illustration of bees collecting honey in a moorland setting, on honey which is not from moorland. Any honey produced outside the UK but which is in a container having a UK name or mark, should be accompanied by a conspicuous indication of the country in which the honey was produced. Blends of honey from two or more countries, which may or may not include the UK, may instead be accompanied by a conspicuous indication that it was the produce of more than one country.

Glossary

Acarine   A disease caused by the mite *Acarapis woodi*, breeding in main thoracic tracheae (breathing tubes).

Acetic acid   Corrosive liquid, used at 80 per cent concentration for sterilising combs, especially against nosema disease spores.

AFB   American foul brood. A very serious brood disease, causing death of the larvae after the cells have been capped. It is caused by an organism called *Bacillus larvae*.

Apiary   Group of two or more hives of bees, called a 'bee yard' in the USA.

Artificial swarm   This consists of one or two combs of bees plus queen hived in a new box on the old site, with the former brood box plus remaining combs of bees removed to another site. The flying bees rejoin the queen, thus producing in effect a swarm. Another queen is then raised in the old box.

Bait hive   A brood box containing one or more old brood combs (to which bees are attracted), for the purpose of enticing a swarm to enter naturally.

Bee space   Room needed by bees to circulate freely within the hive, and left unblocked; usually taken as just over a quarter of an inch (7mm).

Benzaldehyde   Synthetic oil of almonds, used as a bee repellent to clear supers before removal for extraction.

Brood   Term used to describe eggs, open and sealed larvae/pupae in combs; the hive nursery.

Burr comb   Small pieces of thick wax comb used to fill small spaces or just deposited on frames or cover board; called 'brace comb' when linked to other combs or frames.

Glossary

Caramelisation Darkening of honey due to over-heating, usually also noticed by a slight taste of burnt sugar.

Carbohydrate Food substance containing only carbon, oxygen and hydrogen, the latter in the same proportion as water. This provides energy and warmth but is not used for tissue growth.

Carbolic acid Disinfectant used at 8 per cent strength as a bee repellent.

Cast Second or subsequent swarm, having a virgin queen.

Cleansing flight Flight made for the purpose of excreting waste products. This can cause trouble, for example, to a neighbour's washing on the line.

Cluster Group of bees clinging together, as in a swarm, or in winter to conserve heat.

Colony A complete unit of bees, whether in a hive or hollow tree.

Commercial Type of hive using large frames with short lugs.

Crown board Inner cover, normally of thin wood, and framed to create a bee space when placed on the top-most box. It usually has one or two slots for feeding or insertion of Porter escapes.

Culling Elimination of unfit or weaker queens or stocks.

Dadant Hive with very large frames having short lugs; often used by commercial beekeepers in Britain.

EFB European foul brood. A serious brood disease, causing death while the cells are still unsealed. An organism called *Mellissa coccus pluton* (formerly called *S. pluton*) is responsible.

Eke A shallow hive box (about three inches deep) used to create extra space. Originally a shallow ring of plaited straw to give extra depth to a skep.

Enzyme Technical name for a substance which enables a chemical reaction to take place more rapidly; an organic catalyst, for example, invertase.

Glossary

Excluder   Slotted or wired screen with gaps of 0.165 inch (4.2mm) permitting worker bees but not drones or queens to pass through.

Foundation   A thin sheet of beeswax embossed with hexagonal cell shapes on both sides, which bees will accept and build up into full depth combs.

Hertzog   Actually the name of a manufacturer (Schramberg, West Germany), best known for a very rigid queen excluder.

HMF   Hydroxymethylfurfuraldehyde, a substance formed in honey when it is over-heated. Its presence is used as an indicator showing that the honey has lost some of its natural goodness.

Hoffman   A type of self-spacing frame with shoulders, in which the bevelled edge of one meets the flat surface of the next.

Honey-gate   Special tap giving a wide opening for a slowly flowing liquid. It can be quickly shut by a sliding plate cover.

Langstroth   Type of hive most popular world-wide, named after its American originator, Revd L. L. Langstroth (1810-1895).

Laying workers   In a colony long queenless, a few worker bees will develop their rudimentary ovaries unnaturally so that they can lay a few eggs, often several in each cell, giving rise to dwarf drones reared in high-domed cells scattered around the combs.

Mandibular glands   Located in the heads of the queen and workers, producing pheromones which affect colony behaviour.

Mini-nuc   A very small nucleus box used for queen mating, after a ripe queen cell has been inserted.

National   The most popular type of hive used in Britain.

Nectar   Dilute sugary liquid secreted by flowers to attract bees (and other insects).

Nosema   Infection of the bee gut with minute protozoa *Nosema apis*. Very prevalent, as spores remain active in traces of bee excrement.

Nucleus   Small unit of bees, set up by beekeepers for queen mating or for building up into a full colony by feeding.

Orientation flight   Also known as 'nursery flight' or 'play flight'. Made around midday by young bees exercising their wings and learning the appearance of their hive.

Paradichlorobenzine   White crystals used to repel wax moths from stored supers and wax combs generally. Abbreviated to PDB.

Pheromone   External hormone conveying message to other bees, picked up on their antennae.

Porter escape   Bee valve permitting a bee to pass through in one direction only, by means of fine springs.

Prime swarm   First or main swarm, having the original mated queen, leaving behind several queen cells from which virgins will emerge, one only mating and taking over as queen.

Propolis   Sticky brown substance obtained from tree buds. Used by bees to block very small holes, and as a natural antibiotic.

Protein   Important food constituent, needed for tissue building. Obtained by bees from pollen.

Pupa cases   Thin skins left lining the cell walls after the young adult has formed; the final moult.

Quilt   A cover cloth made of canvas or some other strong material, sometimes used instead of a crown board to cover the frames in the top-most box.

Radial   Type of centrifugal extractor in which the uncapped honey-combs are placed like spokes in a wheel, with honey spun out of both faces at once.

Rampin   Tool used instead of a hammer to push in small nails; especially useful in the construction of comb frames.

Ripening   Process of concentrating nectar or sugar syrup by fanning, to evaporate excess water. At the same time the bees add an enzyme (invertase) from their glands to transform any sucrose (ordinary sugar) into a mixture of fructose and glucose (simple, or fruit sugars).

Robbing The stealing of food, usually by other bees, but also sometimes by wasps. A weak colony is vulnerable to robbing, especially if the hive has a full-width entrance, and above all if being fed by syrup. It can result in the loss of the colony.

Scout bees Foraging bees which look for a site suitable for the swarm to occupy.

Skep Straw basket old-fashioned hive, now mostly used just for taking swarms.

Stock Complete unit of bees, that is, the hive and furnishings plus the bees.

Supersedure Process whereby a young queen takes over from her mother without swarming, usually peacefully with both queens laying at the same time for a while.

Tangential Type of centrifugal extractor in which the uncapped honeycombs are placed around the circumference of the rotating cage, so that although only one side is extracted at a time, the combs are drained more completely.

Tanging Noise made by banging on pots and kettles, which was reputed to make a flying swarm settle.

Uncapping Process of slicing off the wax cappings of cells from combs to expose the honey before placing in the extractor.

Waldron Type of queen excluder with parallel metal wires framed in wood.

Wax glands Four pairs of glands situated on the under-surface of the worker bee's abdomen, secreting wax platelets when worker is gorged with food.

WBC Hive invented by William Broughton Carr, using telescopic outer cover lifts to protect the inner bee boxes; of a very attractive appearance.

Varroa Paraistic mite which attacks honey bees and builds up in number year by year, often killing the colony.

Conversion Tables

WEIGHT

kg	lb/kg	lb
0.23	0.5	1.1
0.45	1	2.21
0.91	2	4.41
1.36	3	6.61
3.63	8	17.64
4.54	10	22.05
5.45	12	26.52
11.35	25	55.13
12.71	28	61.77
13.62	30	66.15
22.7	50	110.25
22.42	56	123.48
27.42	60	132.3
45.4	100	220.5

oz	g
1	28.35
2	56.7
3	85.05
4	113.4
6	170.1
8	226.72
12	340.2

LENGTH

mm	inch	cm
6.35	0.25	0.63
9	0.35	0.9
10	0.394	1
12.7	0.5	1.27
25.4	1	2.54
76.2	3	7.62
152.4	6	15.24
457.2	18	45.72

VOLUME

pint	litre/pint	litre
1.76	1	0.57
5.28	3	1.7
7.04	4	2.27
14.08	8	4.54

litre	gallon/litre	gallon
4.55	1	0.22
18.18	4	0.88

Useful Addresses

MAGAZINES

British Beekeepers' Association News Mrs S. Blake, Stratton Court, Over Stratton, South Petherton, Somerset TA13 5LQ.

Bee Craft Mrs S. White, 15 West Way, Copthorne Bank, West Sussex RH10 3QS.

Beekeeping Brian Gant, Leat Orchard, Grange Road, Buckfast, Devon TQ11 0EH.

British Bee Journal C. Tonsley, 46 Queen Street, Geddington, Kettering, Northamptonshire NN14 1AZ.

Bee Quarterly J. Phipps, Walnut Cottage, Pilham, Gainsborough, Lincolnshire DN21 3NU.

Irish Beekeeper Graham Hall, Weston, 38 Elton Park, Sandycove, Co. Dublin, Ireland.

Scottish Beekeeper D. Blair, 44 Dalhousie Road, Kilbarchan, Renfrew PA10 2AT.

ASSOCIATIONS AND SUPPLIERS

Bee Farmers' Association
B. Stenhouse, Borders Honey Farm, Newcastleton, Roxburghshire TD9 0SG.

British Beekeepers Association
Exam Secretary, Charnwood, Beechfield, Barlaston, Stoke-on-Trent, Staffordshire.

Useful Addresses

British Beekeepers Association
General Secretary, National Beekeeping Centre, N.A.C. Stoneleigh, Warwickshire CV8 2LZ.

Bee Diseases Insurance Limited
J.D. Frimstone, 8 Tower Road, Heswell, Wirral L60 6RT.

British Isles Bee Breeders' Association
A. Knight, 11 Thomson Drive, Codnor, Ripley, Derby DE5 9RU.

Candle Makers Supplies
28 Blyth Road, London W4.

Central Association of Beekeepers
Mrs M.R. English, 6 Oxford Road, Teddington, Middlesex TW11 0PZ.

Exeter Bee Supplies
Exeter Road Industrial Estate, Okehampton, Devon, EX20 1UD.

International Bee Research Association
18 North Road, Cardiff CF1 3DY.

Kemble Bee Supplies
Brede Valley Bee Farm, Cottage Lane, Westfield TN35 4RT.

Ministry of Agriculture
ADAS National Beekeeping Unit, York.

Protective Clothing
Sherriff, Five Pines, Mylor Downs, Falmouth, Cornwall TR11 5UN.

Steele & Brodie
Beehive Works, Wormit, Fife DD6 8PG.

Thorne Ltd. Beehive Works
Wragby, Lincoln LN3 5LA.

Welsh Beekeeping Centre
Coleg Howell Harris, Brecon, Powys LD3 9SR.

Index

Index